D U

BREEDING, REARING and MANAGEMENT
By
REGINALD APPLEYARD

SECOND EDITION

Copyright © 2013 Read Books Ltd.
This book is copyright and may not be
reproduced or copied in any way without
the express permission of the publisher in writing

British Library Cataloguing-in-Publication Data
A catalogue record for this book is available from the
British Library

Poultry Farming

Poultry farming is the raising of domesticated birds such as chickens, turkeys, ducks, and geese, for the purpose of farming meat or eggs for food. Poultry are farmed in great numbers with chickens being the most numerous. More than 50 billion chickens are raised annually as a source of food, for both their meat and their eggs. Chickens raised for eggs are usually called 'layers' while chickens raised for meat are often called 'broilers'. In total, the UK alone consumes over 29 million eggs per day

According to the Worldwatch Institute, 74% of the world's poultry meat, and 68% of eggs are produced in ways that are described as 'intensive'. One alternative to intensive poultry farming is free-range farming using much lower stocking densities. This type of farming allows chickens to roam freely for a period of the day, although they are usually confined in sheds at night to protect them from predators or kept indoors if the weather is particularly bad. In the UK, the Department for Environment, Food and Rural Affairs (Defra) states that a free-range chicken must have day-time access to open-air runs during at least half of its life. Thankfully, free-range farming of egg-laying hens is increasing its share of the market. Defra figures indicate that 45% of eggs produced in the UK throughout 2010 were free-range, 5% were produced in barn systems and 50% from cages. This compares with 41% being free-range in 2009.

Despite this increase, unfortunately most birds are still reared and bred in 'intensive' conditions. Commercial hens usually begin laying eggs at 16–20 weeks of age, although production gradually declines soon after from approximately 25 weeks of age. This means that in many countries, by approximately 72 weeks of age, flocks are considered economically unviable and are slaughtered after approximately 12 months of egg production. This is despite the fact that chickens will naturally live for 6 or more years. In some countries, hens are 'force molted' to re-invigorate egg-laying. This practice is performed on a large commercial scale by artificially provoking a complete flock of hens to molt simultaneously. This is usually achieved by withdrawal of feed for 7-14 days which has the effect of allowing the hen's reproductive tracts to regress and rejuvenate. After a molt, the hen's production rate usually peaks slightly below the previous peak rate and egg quality is improved. In the UK, the Department for Environment, Food and Rural Affairs states 'In no circumstances may birds be induced to moult by withholding feed and water.' Sadly, this is not the case in all countries however.

Other practices in chicken farming include 'beak trimming', this involves cutting the hen's beak when they are born, to reduce the damaging effects of aggression, feather pecking and cannibalism. Scientific studies have shown that such practices are likely to cause both acute and chronic pain though, as the beak is a complex, functional organ with an extensive nervous supply. Behavioural evidence of pain after beak trimming in layer hen chicks has been based on the observed reduction in pecking behaviour, reduced activity and social behaviour, and increased sleep duration. Modern egg laying breeds also frequently suffer from osteoporosis which results in the chicken's skeletal system being weakened. During egg production, large amounts of calcium are transferred from bones to create egg-shell. Although dietary calcium levels are adequate, absorption of dietary calcium is not always sufficient, given the intensity of production, to fully replenish bone calcium. This can lead to increases in bone breakages, particularly when the hens are being removed from cages at the end of laying.

The majority of hens in many countries are reared in battery cages, although the European Union Council Directive 1999/74/EC has banned the conventional battery cage in EU states from January 2012. These are small cages, usually made of metal in modern systems, housing 3 to 8 hens. The walls are made of either solid metal or mesh, and the floor is sloped wire mesh to allow the faeces to drop through and eggs to roll onto an egg-collecting conveyor belt. Water is usually provided by overhead nipple systems, and food in a trough along the front of the cage replenished at regular intervals by a mechanical chain. The cages are arranged in long rows as multiple tiers, often with cages back-to-back (hence the term 'battery cage'). Within a single shed, there may be several floors contain battery cages meaning that a single shed may contain many tens of thousands of hens. In response to tightened legislation, development of prototype commercial furnished cage systems began in the 1980s. Furnished cages, sometimes called 'enriched' or 'modified' cages, are cages for egg laying hens which have been designed to overcome some of the welfare concerns of battery cages whilst retaining their economic and husbandry advantages, and also provide some of the welfare advantages of non-cage systems.

Many design features of furnished cages have been incorporated because research in animal welfare science has shown them to be of benefit to the hens. In the UK, the Defra 'Code for the Welfare of Laying Hens' states furnished cages should provide at least 750 cm^2 of cage area per hen, 600 cm^2 of which should be usable; the height of the cage other than that above the usable area should be at least 20 cm at every point and no cage should have a total area that is less than 2000 cm^2. In addition, furnished cages should provide a nest, litter such that pecking and scratching are possible, appropriate perches allowing at least 15 cm per hen, a claw-shortening device, and a feed trough which may be used without restriction providing 12 cm per hen. The practice of chicken farming continues to be a much debated area, and it is hoped that in this increasingly globalised and environmentally aware age, the inhumane side of chicken farming will cease. There are many thousands of chicken farms (and individual keepers) that treat their chickens with the requisite care and attention, and thankfully, these numbers are increasing.

CONTENTS

	PAGE
FOREWORD	5
ADVANTAGES OF KEEPING DUCKS	7
MAKING A START	13
BREEDS AND VARIETIES	16
HOUSING ADULT STOCK	61
FOODS AND FEEDING	69
THE WATER SUPPLY	81
SELECTING BREEDING STOCK	88
TRAP-NESTING	91
EGG COLLECTION AND INCUBATION	94
REARING METHODS	108
MARKING AND RINGING	118
DISTINGUISHING THE SEXES	122
DISEASES AND AILMENTS	126
KILLING AND PLUCKING	131
ORNAMENTAL WATERFOWL	133
SHOWING	140
NOTES FOR NOVICES	145
GLOSSARY OF DUCK-KEEPING TERMS	148

LIST OF ILLUSTRATIONS

	PAGE
COMMERCIAL DUCK KEEPING	FRONTISPIECE
UTILITY PEKINS	9
LAY-OUT FOR A DUCK FARM'S ADULT STOCK	11
ORCHARD RUNS	14
AYLESBURY DRAKE	18
PEKIN DUCK	21
ROUEN DUCK	23
BUFF ORPINGTON BREEDING QUARTETTE	29
FLOCK OF BLACK AND WHITE MAGPIES	33
CAYUGA DRAKE	35
WHITE AND FAWN RUNNERS	40
UTILITY KHAKI CAMPBELLS	49
WHITE CAMPBELLS	51
MUSCOVY BREEDING PEN	53
WHITE CRESTED DUCKS	56
TYPICAL UTILITY CROSS	59
VERMIN-PROOF NIGHT SHELTER	62
AN ECONOMICAL SHELTER HOUSE	63
HOME-MADE STRAW SHELTER	65
SHELTER FOR LAYING COMPOUND	67
APEX DUCK HOUSE	68
MASH MIXING TABLE	76
DUCKLING DRINKING TROUGHS	77
DUCKS IN NATURAL SURROUNDINGS	79
CEMENT POOL	82
WATERING ATTACHMENT FOR INTENSIVE HOUSE	86
TRAP-NEST BOXES	91
TRAP-NEST FRONTS	92
SUN PARLOUR DUCKLINGS	93
IMPROVISED BROODY NESTS	98
TESTING STAND	104
PHASES IN INCUBATION	105
COMBINATION HOUSE AND BROODER	111
COMMERCIAL BROODING HOUSE	113
TOE MARKING	119
CAROLINA DRAKE	135
DUCK COMPOUND	137
CEMENT POOL FOR ORNAMENTAL DUCKS	138
SHOW TYPE KHAKI CAMPBELL DRAKE	142
SHOW TYPE KHAKI CAMPBELL DUCK	143

FOREWORD

HAVING kept and bred many breeds and varieties of livestock—mice, cagebirds, ducks, poultry, geese, sheep, pigs, goats, cattle and horses—I can honestly say that I know of no livestock which can prove of more interest or give so much pleasure as ducks.

Truly they have intelligence and use their brain. They have given me joy and much pleasure over many years and through them I have made many friends.

In this book I have dealt with the points that strike me as likely to be most helpful to the beginner—and I hope to the experienced also—in the breeding and rearing of ducks, their management, etc.

I have in mind many pitfalls, most of which I have fallen into at one time or other in my many years' experience, and it is hoped to help the reader to avoid these expensive slips. They, and the disappointments that go with them, are easily avoided, if one knows how ! Readers must read between the lines, and use his or her own judgment in applying the advice given as thought best.

Duck production is attracting newcomers to its ranks every day, one of the greatest points in its favour being the comparative freedom of ducks from disease. That being so the profits from this branch of the Industry, embracing both eggs and table ducks, are appreciably higher than with fowls.

REGINALD APPLEYARD.

Ixworth,
Suffolk.

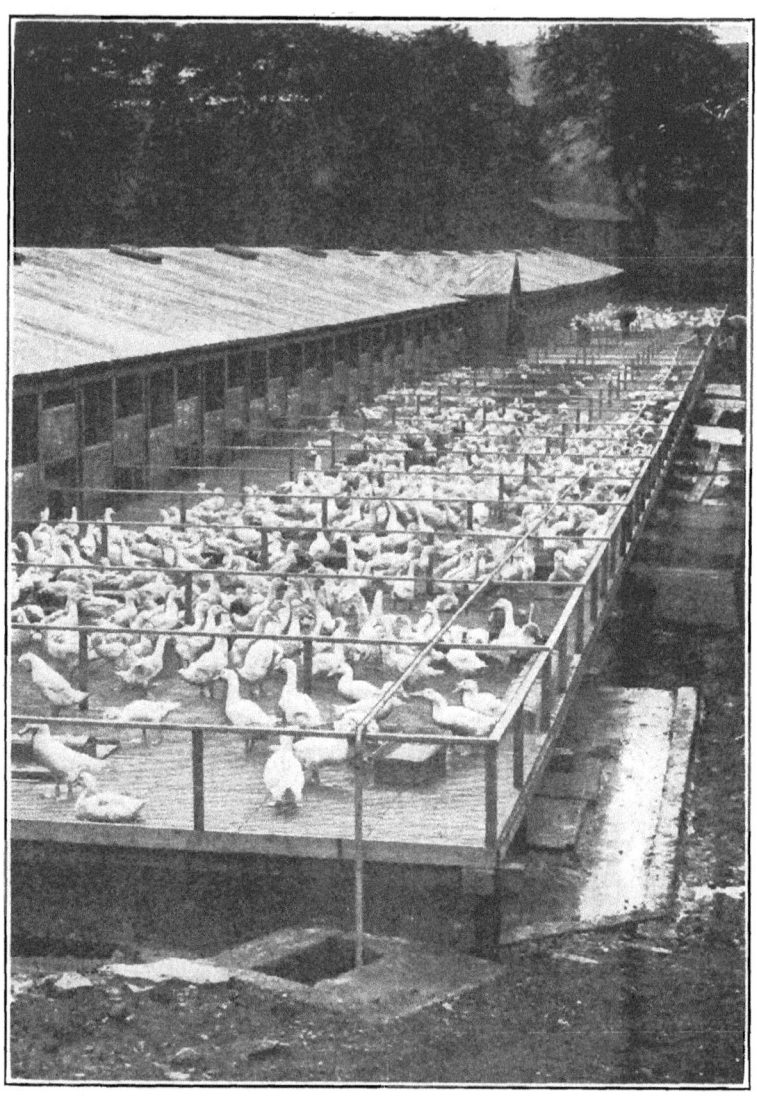

Frontispiece

The provision of all the year round ducklings calls for an extensive use of the system of housing such as is depicted in this photograph. The ducklings are, of course, kept intensively

DUCKS

Chapter I

ADVANTAGES OF KEEPING DUCKS

IN the domesticated duck we have a most useful, sensible and brainy bird, one which can be kept successfully and made profitable even under adverse conditions. By adverse conditions I mean that many breeds of duck can be kept on dry land with success so long as they are given a plentiful supply of clean water in suitable containers.

In most cases when keeping such as Khaki Campbells, Indian Runners and other active breeds of laying ducks, also Pekins and Aylesburies purely as table birds, it can be said that they often do just as well and give as good results as those birds enjoying natural water in ponds, streams and rivers.

In some cases it has been proved that flocks of ducks with only supplied water have done better than others with natural water. This can be put down to the fact that ducks are water lovers; even in very cold, bad weather they will spend much of their time on the water and this is said by many to be against the egg output, whereas ducks with only water in containers do not spend so much time in cold water and thus can conserve more heat energy to go to the making of eggs.

The main essential to success with ducks, especially as egg producers, is to have them tame, happy and contented both with their attendant and their surroundings. I would put this even before special and elaborate feeding and housing. The wild, scared duck cannot make a success in life. On the other hand, the duck which is bred and reared right from a highly fecund strain, properly cared for and fed with sensible foods, talks to its attendant and thus shows contentment, will give both pleasure and profit to its owner. Contentment leads to success in the duck and by success I mean plenty of eggs from laying breeds, and lots of fertile eggs from the table breeding stock, which will mean many fat and contented table ducklings.

There are several advantages to be had from keeping a few or many ducks; as a profitable sideline and a grand hobby, they are great on any farm or small holding. They are hardy and real money earners. One big advantage is that the duck does not ask for or require any elaborate housing with a multitude of windows, droppings-boards, perches, nest boxes, etc. Very much the opposite; all that is necessary is a roomy, low, well-ventilated house, or some old building converted to their requirements. It is even better if the houses have not got lots of windows, as on light and moonlight nights they only throw shadows which scare the ducks.

It is both possible and practical to keep laying and breeding stock in the open, without the expense of housing. This I propose to deal with in detail later on. Ducks are profit earners; they are not subject to untold diseases, and they have a long and profitable life, from a commercial point of view, of 2 to 2½ and often 3 years. As breeders they will live up to 15 or more years when leading a natural life.

I once remarked to a friend, " All your adult Black East Indians are black and white—why? " " Old age " came the reply, " I brought them with me when I came here, and that was 15 years ago ! " In this case all the old ducks had laid and each had reared youngsters; in one case eleven and in another eight, and all strong ducklings. Also, many eggs had been taken early in season for eating and cooking. Proof that ducks are hardy and long lived.

A good point is that ducks lay 98 per cent. of their eggs before 9 a.m.—not all over the place, but in their house or compound.

Another good feature is that laying breeds and heavy breeds to be kept as future stock need not be hatched before mid-

DUCKS

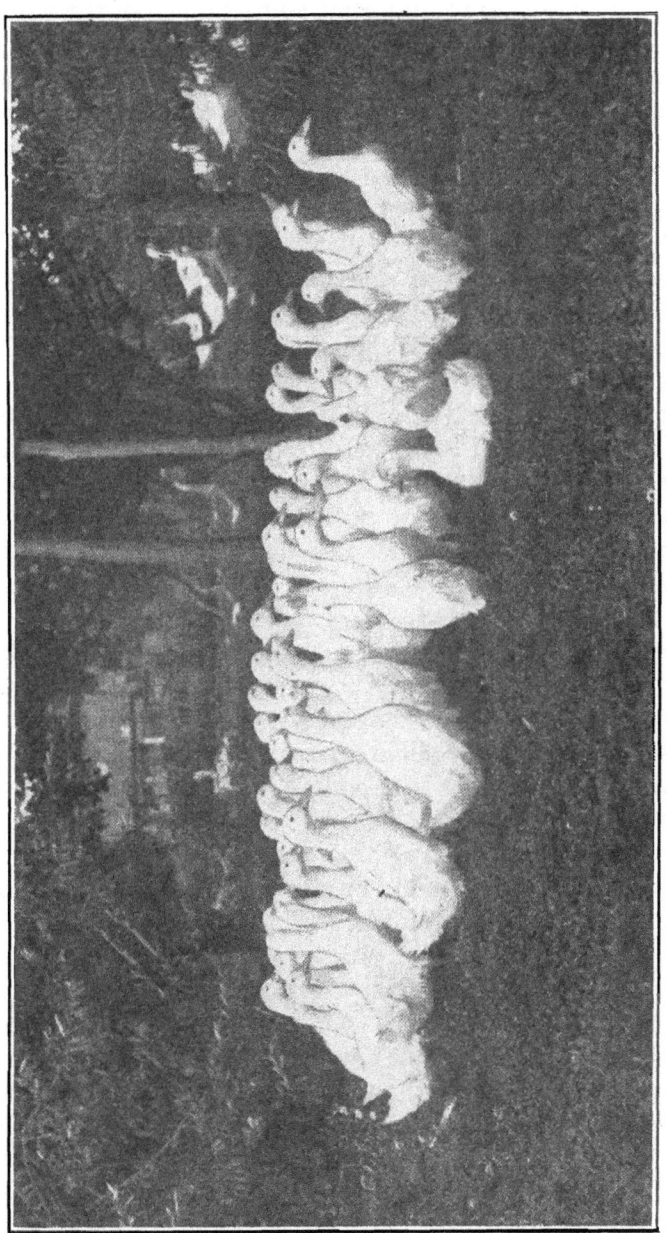

GROUP OF UTILITY WHITE PEKINS
This breed makes excellent weights quickly and is ready for killing at from 9 to 12 weeks

April and onwards up to mid-May, so that much of the worry of early hatching and rearing in bad weather is cut out.

The reader may well say, " Well, after all these good points, there must be some snags somewhere ! " Granted, there are, and one is that the duck is not a back-yarder's bird and must not be kept in a very confined space—except the very small ornamental varieties.

But it can truthfully be said that there are few if any really important snags so long as one has the right class of stock and provided you give it proper care and attention. Compared with most live stock the duck requires quite a small amount of attention, nor is it a dainty, fanciful feeder ; all it asks for is one small feed of grain each morning and one good feed of wet mash each evening, about one hour before dusk.

Have you a suitable place to keep ducks ? If you do keep them, can you find a use or sale for their eggs ? Or a sale for prime table ducks if you take up that side of the industry ? There is a big demand for both—a continual demand if you supply the perfect article.

Have you a small paddock or orchard ? Or some low, marshy land with streams and ponds which at a certain season of the year abound with natural foods in the form of duckweed, water weeds, and many different sorts of insect life—all of great value to the duck as a natural aid to the production of eggs ?

Throughout the country there are thousands of acres of marshy land, next to unused yet ideal for waterfowl. If you have such a property, give the question of stocking it with ducks your careful consideration. Remember, it is not necessary to replace your laying flock each year. With ducks you can, if you wish, hatch your replacements every other year and keep the birds three seasons.

In the case of those owning a small orchard or paddock near the garden and house, remember that ducks cannot scratch ! Wire netting, 3 feet high, 2-inch mesh (or larger) will confine the birds and keep them where desired, and even if they did get into the garden they would only search for

DUCKS

snails, slugs, etc., and do little if any damage. If you care to pinion all the ducklings hatched, 18-inch netting will control them. (See the chapter on "Pinioning and Wing Clipping.")

After using wire netting of every size and shape I am of the opinion that it pays every time to buy the best—never using a larger mesh than 2 inches—larger sized mesh netting is weak and difficult to handle and erect.

In an orchard, ducks will clear up much of the insect life which abounds and at the same time manure the ground and receive shade and shelter from the trees. If you have no

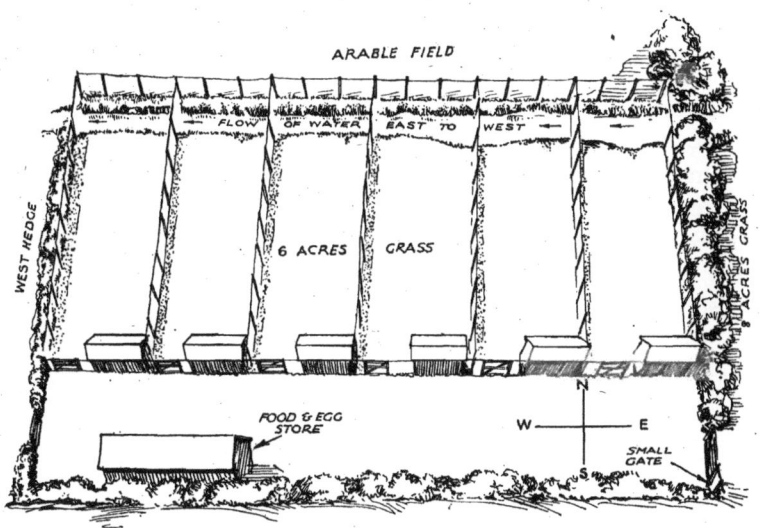

A suggested lay-out for a duck farm's adult laying and breeding flocks

swimming water for the birds, arrange a cement pool for them. (See the chapter on "The Water Supply.")

Have you some low grass land which is infested with liver fluke, upon which you only turn cattle in fear of annual loss? If so, try a flock of ducks; throughout the year they will quickly clear out all the snails which are the host of this pest, and at the same time pay you well as egg producers.

Have you a pond which becomes green each season, or a stream which becomes overgrown with weeds and silted up with mud? If so, try a flock of ducks. From it they will get health and food, produce eggs and at the same time devour the weeds and move the mud, all at no expense in labour, etc.

One special advantage with ducks, in fact with all waterfowl, is that they are very easy to train to routine habits. Properly trained they will return each evening to their house or pen.

They will travel, under their own steam, to their feeding ground quite a distance away, working and finding foods over a large area and returning. This is a great advantage as it means their house or pen can be near at home and so save work and travel for the attendant. Best of all the house or pen can be on a chosen dry spot and the birds travel and work over acres of low land, water meadows, streams and waste land abundant with natural and perfect duck foods.

On very wild overgrown ground it will be found better to keep white birds or a light coloured breed—much easier to see if necessary to round them up in the dusk evenings, also they are less likely to get shot!

Before concluding this chapter let me repeat that the duck is not a backyarder's fowl. I know that many hundreds have been kept with moderate success under restricted conditions, but in fairness to the ducks themselves they are natural foragers and roamers and will quickly turn the small section of the garden handed over to them into a mud-bath.

On range they are not subjected to the upsets that often occur in restricted quarters; even the most enthusiastic breeder will agree that ducks are temperamental and sudden disturbances will often bring about a premature moult.

Chapter II

MAKING A START

IT is only too easy to rush into anything, spend good money and then find you have got the wrong thing. In this short chapter I propose to put myself in the place of one who has suitable ground and wishes to commence in ducks.

How shall I make a start? Much depends on the season of the year. If in the autumn, I can start by the purchase of a trio, pen or even a small flock of correctly mated breeding stock. Depending on the breed, it should be possible to buy reliable breeding birds. It is better to acquire a few tip-top birds likely to breed perfect progeny than to rush in and purchase lots of inferior birds from all over the place.

A trio—a drake and two ducks, properly mated, first- or second-season birds, would last as foundation stock for a number of years and if used purely as breeders and not forced I would expect each season to rear from them a number of layers, the choicest of which would be breeders for the future.

Even if I bought stock, that would not prevent me from buying a few sittings of eggs at the spring of the year, toe-marking the resulting ducklings and maybe mating some of them up later on for the future.

When purchasing eggs you can generally take your choice of 12 eggs and " clears " replaced or 15 eggs and no replacements. Usually if 12 eggs are taken the seller will replace any " clear " eggs which are returned to him for examination, say, within 14 days.

A " clear " egg is one which has no signs of incubation and which when held before a light looks just like a fresh egg, except that the air space at the top will be larger. Such eggs will be replaced.

Eggs which have been fertilized by the male and then the

germ dies are not replaced. If you go to a reliable breeder and fancier, who values his name and reputation, my advice is take 15 eggs, one trouble in setting and in rearing the results. Most breeders will meet your request for less number of eggs or even mixed sittings from different breeds and varieties.

By going to a reliable breeder and buying birds, however, I would certainly have something to look at and to carry on with for the future; better still, they would be correctly mated and likely to breed good birds. If they did so I would

Ducks will find much natural food and will enjoy the shade afforded by the trees if allowed to run in an orchard. They will not damage bush fruit

leave the original pen for a number of years to work up into a large flock.

Alternatively, ducklings could be purchased, so there are a number of ways of making a start, depending on the season of the year and circumstances.

A drake and three, four or five ducks would give me more eggs to sit at the right season of the year. However, in the case of foundation stock I would increase the period of hatching, toe-marking and ringing all the progeny. The later ones could be run on, say, as layers and later, when in the second season, they would prove reliable breeders.

With all these alternatives it is up to the would-be duck-keeper to use his own ideas and judgment as to the best method of starting. Make out a plan and go right out to carry it through to success. Just one other thing—try to make up your mind once and for all on the breed you wish to keep, do not keep chopping and changing, it costs good money and wastes time. Get on to a good breed which you like and which is suitable for your purpose and surroundings and stick to it. Get to know all you can about the breed and try to improve it for yourself and for the good of the breed.

As later chapters will show, there are other egg-laying ducks than Khaki Campbells and Indian Runners, while the Aylesbury need not necessarily be regarded as the only table type of duck.

I like the beginner who sets himself a programme and sticks to it. He is, in fact, the one type of person likely to succeed. Failures are usually found among those who continually chop and change and never appear to be satisfied with what they are doing.

Chapter III
BREEDS AND VARIETIES

THERE are many useful and beautiful breeds of ducks; a variety to suit most purposes, situation and taste as to colour, shape and size.

Breeds suitable for producing purely table ducks are Aylesburys or Pekins. Others, which are layers, such as Magpies, Buff Orpingtons, Blue and Black Orpingtons, Stanbridge Whites and large utility Fawn and White Indian Runners are also useful for table purposes.

The Campbell White and Khaki all give drakes, and ducks, which are quite useful eating—delicious in flavour and with good deep breast meat. To my findings they have much deeper breast fleshing than the purely table breeds and best of all are light in bone.

For those who desire birds solely for ornament on large or small waters, the case might be met with such as Teal, Widgeon, Pintail, Shoveller, Pochard, Tufted, Carolina, Mandarin and other extraordinarily beautiful waterfowl.

Then we have Rouens, Cayugas, Muscovies, Crested, etc., which are very beautiful, and any birds not required are excellent for the table.

As a small, very hardy, ornamental breed there is the Black East Indian, and the Brown and the White Decoy.

For those who take up any breed and wish for full and further information, I would advise them to become a member of the breed club and join the British Duck Keepers' Association.

It might be well to mention and pass on the advice given to me by one of our best and oldest livestock breeders—a man of great practical experience—" My advice to you, my boy, is never make the mistake of thinking you can beat nature, you can help and aid nature in some cases, but

DUCKS

never go on struggling to keep and breed birds, or for that matter, any stock which does not suit your district and climate. Some breeds must have climate and surroundings suitable to them. . . ."

I have found this to be perfectly true, yet I still struggle with certain breeds although I know the advice received is practical and sound and was given free !

It may be the land or the water—maybe the climate or cold east winds—or exposed position.

THE AYLESBURY

This is the best of all table breeds : very large, white-fleshed, getting its name from the vale of Aylesbury where the breed originated, and where it is still reared in vast numbers each year for the London market.

It is only fair to say that there are many white ducks called Aylesburys which are most decidedly not Aylesburys either in type or breed characteristics ; simply mongrel white ducks with just a little pure Aylesbury blood.

The true Aylesbury is just a moderate layer, and if of a good strain it will generally give a flock average of a hundred eggs or so in the year. Remember, they are bred for meat, not high egg records.

Flocks of Aylesburys generally lay early in the New Year and continue to the end of June or early July, when they go into a moult and out of production. With careful management and good feeding it is possible to get some eggs in the early autumn months, but most breeders prefer to rest their birds, get them into production early in the New Year, and produce the much desired early spring duckling, which obtains such a good price per pound.

The chief difficulty is unfertile eggs early in the year and early spring ; thus, great care should be given to the choice of really active, sound drakes, known to be bred from extra fertile stock.

With the very large " keely " type of stock a useful mating

is a drake with two or three big ducks. With the lighter utility types, using a more active drake, as many as five and in some cases six females may be given—or they may be run as a small flock, two drakes with ten or twelve ducks.

A typical Aylesbury is best described by giving The Poultry Club Standard of the breed.

General Characteristics. HEAD.—Large, straight and long. BILL.—Long and broad, and when viewed from the side, the outline almost straight from the top of the skull, the head and

AYLESBURY DRAKE
Universally recognised as the leader among the table varieties of ducks

bill measuring 6 to 8 inches. EYE.—Full. NECK.—Long, slender and slightly curved.

BODY.—Long, broad and very deep; full and prominent breast. KEEL.—Quite straight from breast to stern; straight back, almost flat; strong wings carried closely to the sides, fairly high but not touching across the saddle.

TAIL.—Short, only slightly elevated, and composed of stiff feathers, the drakes having two or three curled feathers in the centre.

LEGS.—Very strong and short, the bones thick, well set to balance the body. TOES.—Straight, connected by web.

CARRIAGE.—Horizontal, the keel practically parallel with the ground.

PLUMAGE.—Bright and glossy, resembling satin.

WEIGHT.—Drake 10 lbs., Duck 9 lbs.

COLOUR. BILL.—Pink, white or flesh. EYE.—Dark. LEGS and FEET.—Bright orange. PLUMAGE.—Pure white.

SCALE OF POINTS

Head and Bill 20, Eyes 8	28
Size	20
Condition	12
Keel	10
Type	10
Colour	10
Neck	5
Legs and Feet	5

Serious Defects. Plumage other than white; bill other than white or flesh-pink; any deformity; ducks very heavy behind.

THE PEKIN

To the average man in the street any old white duck of good size is an Aylesbury or perhaps a Pekin! Really there is a big difference, just as much difference as between chalk and cheese, at least in appearance.

There are two distinct types of Pekins. As now bred in England and America for table duckling production, the

Pekin is white in colour, but still with orange bill, legs and feet. The idea is to breed out the yellow pigment both in flesh and feather, and to produce a Pekin in type without this undesirable yellow colour.

The original Pekin was a yellow, or creamy colour; and a really good exhibition bred bird should still be this colour. To be plain, if you wish to exhibit and win in this variety, you must have both size and the desired yellow feather coloration.

However, from a table duckling point of view this yellow pigment is undesirable and of no value; in fact it is most detrimental to the value of the plucked duckling when offered to the salesman or shop. Why this should be I am at a loss to understand, but it is true nevertheless.

Breeders have now made great headway in breeding out this yellow colour, and we have many strains in the country which are Pekin in type but with white feathering and white skin. They are big, fine birds with great length, width and depth of body.

There are also many good strains of purely exhibition Pekins for those who wish to show. A good Pekin will always get into the money in " Any Other Variety " classes. It is said they were first imported into England in 1873, and first shown in 1874.

The Pekin is definitely a fine table bird, large, naturally well fleshed and plump, matures quickly, and carries flesh in the right place. The commercial strains are extra fertile, and the ducks very good layers of pure white eggs of beautiful shell texture; this alone is worth a lot, as a good-shelled egg hatches well.

The Pekin is a useful breed to cross with the Aylesbury, giving strong, vigorous, easily reared ducklings.

It will be noted from The Poultry Club Standard which follows that the Pekin is ideal as to size, and type for table duckling production if the yellow colour is done away with. By proper methods and feeding it is possible to breed and place on the market a duckling with little, if any, yellow tinge.

DUCKS

I can honestly say that if I wished to make a living by producing table ducklings, it would be with Pekins, and if the yellow orange bill and some slight yellow skin coloration persisted I should expect to be well covered by the extra percentage of ducklings reared to good saleable age for the number of eggs incubated. Also, I should expect to get a good price for white coloured eggs sold when not required for incubation.

PEKIN DUCK
One of the leading table breeds either in a pure state or when crossed

Working with big utility White Pekins it is found that the males are generally excellent sires and very fertile, the usual mating being five and six ducks per drake. In the case of the purely exhibition yellow plumaged birds fertility is not so good and usually they are mated as trios, or a drake given three special ducks as a limit.

General Characteristics. HEAD.—Large, broad and round,

with high skull, rising rather abruptly from the base of the bill, and heavy cheeks. BILL.—Short, broad and thick, slightly convex but not dished. EYES.—Partially shaded by heavy eyebrows and bulky cheeks. NECK.—Long and thick, carried well forward in a graceful arch or curve, and with slightly gulleted throat.

BODY.—Broad and of medium length, and without any indication of keel, except a little between the legs; broad, full, followed in underline by the keel (which shows very slightly between the legs) to a broad, deep paunch and stern, carried just clear of the ground. Broad back, short wings carried closely to the sides; well-spread tail carried high, the drakes having two or three curled feathers on top. (*Note.*—A good description of the general shape of the Pekin is that it resembles a small wide boat standing almost on its stern, and the bow slightly forward.)

LEGS.—Strong and stout, set well back and causing erect carriage. TOES.—Straight, connected by web.

CARRIAGE.—Almost upright, elevated in front, and sloping downward to the rear.

PLUMAGE.—Very abundant, the thighs and fluff well furnished with long, soft, downy feathers.

WEIGHT.—Drake 9 lbs., Duck 8 lbs.

COLOUR. BILL.—Bright orange and free from black marks or spots. EYES.—Dark lead-blue. PLUMAGE.—Buff-canary, sound and uniform, or deep cream, the former preferred.

SCALE OF POINTS

Type	25
Size	20
Head (Bill 10, Eyes 5)	15
Colour	15
Condition	10
Neck	5
Tail	5
Legs and Feet	5

DUCKS

Serious Defects. Black marks or spots on the bill. White plumage. Any deformity.

THE ROUEN

The Rouen can certainly be called the most beautiful and magnificent of all the domesticated ducks. It is of glorious colour and of nice size, often weighing twelve or more pounds as a yearling.

It is not easy to breed a really good exhibition Rouen. Quite the opposite; but this makes them all the more interesting to keep. There are Rouens and Rouens. The purely exhibition type of big size has been bred generation after generation for type and coloration. Such birds are not free breeders and are often very infertile. Working with such birds, I usually put two or three ducks to a

ADULT ROUEN DUCK
Of the approved exhibition type

male. Then there are others more of utility type—not so big, but more active, and these are mated one drake to four or five ducks.

The drake, of course, is a bird of wonderful colour, and so is the duck, with her rich brown or chestnut, and each feather definitely marked with black or very dark brown lacing. The stripes or Mallard marking on her face are also most attractive.

The Rouen has very good table properties; used as a cross with Aylesburys or Pekins, it gives a fine bird. Such crosses are really beautiful, coming all shades of golden and silver dun. A good mating is a drake and two or three ducks.

It is difficult to explain the beautiful colour of the male on paper, but the following Standard will give a good idea.

General Characteristics. HEAD.—Massive. BILL.—Long, wide and flat, set on in a straight line from the top of the eye. EYE.—Bold. NECK.—Long, tapering and erect, slightly curved, but not arched.

BODY.—Long, broad and square, deep keel, just clear of the ground from stem to stern; broad and deep breast; large wings well tucked to the sides; very slightly elevated tail, the drakes having two or three curled feathers on the top in the centre.

LEGS.—Medium length, stout shanks, well set on to balance the body in a straight line. TOES.—Straight, connected by web.

CARRIAGE.—Horizontal, the keel parallel with the ground and just clear of it.

PLUMAGE.—Tight and glossy.

WEIGHT.—Drake, 10 lbs.; duck, 9 lbs.

COLOUR. THE DRAKE. BILL.—Bright green yellow, with black bean at the tip. EYES.—Dark hazel. LEGS and FEET—bright brick red. PLUMAGE, HEAD and NECK.—Rich iridescent green to within about an inch of the shoulders where the ring appears. RING.—Perfectly white and cleanly cut, dividing the neck and breast colours, not encircling the neck, but leaving a small space at the back. BREAST.—Rich claret coming well under, cleanly cut, not running into the body colour, and quite free from white pencilling or chain armour.

CHAIN ARMOUR or FLANK PENCILLING.—Rich blue French grey, well pencilled across with glossy black, perfectly free from white, rust or iron. STERN.—Same as flank, very boldly pencilled, close up to the vent, finishing in an indistinct curved line (perfectly free from white) followed by rich black feathers up to the tail coverts.

TAIL COVERTS.—Black or slate black with brown tinge, with two or three green-black curled feathers in the centre.

BACK and RUMP.—Rich green black from between the shoulders to the rump.

WINGS.—Large coverts: pale clear grey; small coverts: French-grey very finely pencilled; pinion coverts: dark grey or slate black; bars (two, composed of one line of white in the centre of the small coverts): grey, tipped with black, also forming a line at the base of the flight coverts, the latter feathers slate black on the upper side of the quill, and rich iridescent blue on the lower side, each of these feathers tipped with white at the end of the lower side, forming two distinct white bars (the pinion bar being edged with black), with a bold blue ribbon mark between the two, each colour being clear and distinct and making a striking contrast; flights: slate-brown with brown tinge free from white.

THE DUCK. BILL.—Bright orange with black bean at the tip, and with black saddle extending almost to each side, and about two-thirds down towards the tip. EYES.—Dark hazel. LEGS and FEET.—Dull orange brown.

PLUMAGE. HEAD.—Rich brown of a golden, almond or chestnut hue, with a wide brown-black line from the base of the bill to the neck and very bold black lines across the head, above and below the eyes, filled with similar lines. NECK.—The same colour as the head, with a wide brown line at the back from the shoulders, shading to the back of the head.

WINGS.—Bars, two distinct white bars with a bold blue ribbon between, as in the drake; flights, slate-black with a brown tinge, no white.

REMAINDER OF PLUMAGE.—Rich brown of a golden, almond or chestnut hue, of level shade, every feather distinctly pencilled from throat and breast to flank and stern, the markings to be rich black or very dark brown, the black pencilling on the rump having a green lustre.

SCALE OF POINTS
THE DRAKE.

Colour, Breast 10, Bill 5, Neck 5, Ring 5, Chain Armour 5, Back and Rump 5, Wings 5, Stern 5, Tail 5	50
Markings	10
Size	10
Type	10
Condition	10
Head	5
Legs and Feet	5

THE DUCK

Colour.—Ground 15, Bill 10, Head 5, Wings 5, Neck 5	40
Pencilling	20
Size	10
Type	10
Condition	10
Head	5
Legs and Feet	5

Serious Defects.—Leaden bill, no wing bars; white flights; stern, broken; wings down or twisted; any deformity. The drake: black saddle, black bill, or minus ring (on neck). The duck: white or approaching white; ring (on neck).

THE ORPINGTON

The Orpington duck was made and originated by the late William Cook of Orpington. There are five colours, Buff, Blue, Black, Chocolate and White.

The Buff should be buff throughout and the White white throughout, whereas the Blues, Blacks and Chocolates have a white heart-shaped bib on the upper part of the breast.

As a breed the Orpington is quite useful. Of medium size, giving excellent full-breasted and plump ducklings, hardy, vigorous, and quite good layers of excellent-sized and generally pure white eggs. They are most suitable for fanciers and private persons who want a useful yet good-looking bird— ideal, in fact, for the hobbyist.

Frankly, it is not easy to breed a perfect bird in any of the colours. The Standard of Perfection gives one a very good idea of each colour. The White has never really taken on and in my opinion, never will, as it is too much like an ordinary good white duck!

The Blues were originated about 1910 and are certainly a beautiful variety, most interesting in that it is not easy to breed good ones of a soft, even shade of blue throughout and free from white in flights. Nor is it easy to get the white bib distinct and of correct shape. Another thing is that one breeds a number of Whites, Whites splashed with grey feathers, others black and some black and white. With careful mating, however, one can form a strain which will give a good big percentage of Blues.

The Black, of course, comes from the Blues and by using some Cayuga blood one can soon breed good Black Orpingtons, although the white bib is again most difficult.

From a utility point of view the Buff is, of course, the best, as it has been bred for eggs and many ducks of this breed have put up good records at the laying tests. The buff required is quite a light shade and the underfluff and any " stubs " in ducklings do not show so much as in the Blues and Black when killed for table purposes at 11 to 12 weeks old.

Buff Orpington drakes are really decent table ducklings, both in colour and texture of skin and flesh if properly fattened and finished off.

A good Orpington, especially a Blue, will generally get

into the money in " Any Other Variety " classes at the shows. The Buffs usually have classes at most of the big shows and have a strong club with a good number of members. A good mating is a drake and four or five ducks.

THE BUFF ORPINGTON STANDARD

The breed is essentially a triple purpose breed. It combines beauty of form and colour with good table quality and profitable egg production. It is contrary to the best interests of the breed and at variance with the correct interpretation of this Standard to breed for any one of these three qualities at the expense of the others.

THE DUCK. HEAD.—Fine and oval in shape. SKULL.—Narrow. EYE.—Brown iris with blue pupil, set high in the head; large and bold, giving the head a look of alertness and activity. A deep-set, scowling eye is objectionable. BILL.—Proportionate to the head in size, upper mandible straight from bean to base in line with highest point of skull. Colour of bill, orange with dark bean. NECK.—Slender, of moderate length, upright.

BODY.—Long, broad and deep, particularly at the shoulders; free from any sign of keel. When in lay, the duck's abdomen should be nearly touching the ground. The carriage of the body should be slightly elevated at the shoulders, not quite so horizontal as the Aylesbury, but avoiding any tendency to confusion with the upright carriage of the Pekin or Runner.

The back should be perfectly straight in line, the tail being small, compact and rising slightly from the line of the back.

Legs should be of moderate length proportionate to the body of the duck, set well apart, and bright orange-red in colour.

The colour of the duck should be a rich even shade of deep red-buff throughout, free from lacing, barring and pencilling, blue, brown or white feathers. The wings should be the same colour throughout as the rest of the body.

The weight of a matured duck in lay should be about 6 lbs.

Serious Defects.—Any physical deformity, such as twisted

wings, wry tail, humped back, twisted bill, etc. Colour other than buff, *e.g.* white feathers in neck or breast, brown feathers, very heavy lacing, strong light line over eye, green bill.

It is recognised to be difficult to achieve good wing colour, and therefore **pale wings**, though objectionable and to be avoided and penalised, are not a disqualification.

BUFF ORPINGTONS
A drake and three ducks make up a good small-unit breeding pen. The Buff Orpington is one of the best dual-purpose ducks

A matured duck should weigh between 5 lbs. and 7 lbs.

THE DRAKE.—The type and general physical characteristics of the drake should be identical with those of the duck after allowing for sexual differences. The chief sexual differences are as follows : Slightly increased length and weight, curved feathers in tail, longer bill, and colour differences, lack of depth in abdomen. The body colour in drakes should be the same as in ducks, and as level as possible throughout with the following differences : Head and neck, seal brown with bright gloss, but complete absence of beetle green. The seal brown of the drake's neck should terminate in a sharply defined line all the way round the neck. The rump should be reddish brown as free from " blue " as possible.

Common faults to be avoided are blue in rump, pale colour or deep brown under tail, white wing.

Scale of Points

Head and Eye 10
Type (shape and carriage) and size . . 40
Colour 40
Condition 10

Serious defects are grey, silver, or blue head, white feathers in neck, brown secondaries, beetle green on any part, very green bill, and any of the physical defects mentioned against the duck.

A matured drake should weigh between 5 lbs. and $7\frac{1}{2}$ lbs.

A well-bred Buff Orpington should be capable of laying at least 180 eggs, weighing not less than $2\frac{1}{2}$ ozs. each, in 12 months under normal domesticated conditions.

Blue Orpington Standard

General Characteristics. HEAD.—Fine and oval shaped. BILL.—Of moderate length and in a straight line from the skull. EYES.—Bold. NECK.—Fairly long, gracefully curved.

BODY.—Long, broad and deep, full round breast, strong wings carried closely to the sides ; small tail rising gently, the drake's having two or three curled feathers in the centre.

LEGS.—Of medium length, strong and well apart ; straight toes connected by webs.

PLUMAGE.—Light and glossy.

CARRIAGE.—Slightly upright.

WEIGHT.—Drake, 7 lbs. Duck, 6 lbs.

COLOUR. DUCK. BILL.—Blue, a greenish tinge permitted, especially in old birds, but no trace of yellow or orange.

HEAD, NECK, BODY and BREAST.—Blue throughout, free from bronze tint, with a white bib extending from centre of neck about three inches on to the breast, roughly oblong about two inches at the widest part. Clearly defined.

THE DRAKE. HEAD and NECK.—A darker blue. The soundest coloured birds will show some lacing; for the present some white in flights to be permitted, but no white in face, and any eye streak to be severely penalised. EYE.—Black pupil with deep blue iris.

LEGS and FEET.—Dark blue, for the present some orange or yellow allowable if mottled with blue, but all yellow or orange to be passed.

PLUMAGE.—An even shade, the darker the better, with a touch of white on the breast. The head and upper part of drake's neck at least two shades darker than his body colour.

SCALE OF POINTS

Head, Bill, Eye (including colour)	20
Legs and Feet (including colour)	20
Size, Carriage and Shape	30
Colour of Plumage	30

Serious Defects.—Twisted wings, wry tail, any other deformity. White in face, all yellow or orange legs. Colour other than stated.

THE MAGPIE

The Magpie is a most attractive duck, beautiful, hardy and useful. Can be had in Black and White, Blue and White, and Dun and White.

To breed a bird with perfect markings of good, sound colour is not too easy; however, this adds interest to the breed. It is certainly a most useful duck, because it will be noticed that the markings are so placed that when a bird is dead and plucked there are no black or coloured stubs to show on the breast; in other words, they pluck out just as clean as a white duck. On the whole the breed lays well.

Breeders have much improved the type, colour and markings of Magpies, making what was just a good black and white (or other coloured) duck into a well-marked and Standard-shaped bird, at the same time, in most cases, paying due regard to its egg-laying powers.

For those who like something out of the ordinary in appearance, combined with good laying qualities and decent table properties, the Magpie is well worth while. The ducklings are very beautiful and immediately they are dry and nicely fluffed out one can have a very good idea as to how they will be marked when feathered.

In some cases this is useful as those youngsters which are poorly marked can be sold to be finished off for table. A good average mating is a drake with five ducks.

Standard for Magpie Ducks

General Shape and Carriage.—Fairly broad across and deep; great length of body, giving a somewhat racy appearance, indicative of strength combined with great activity.

Head.—Long and straight. Neck.—Long, strong and nicely curved. Bill.—Long, broad and slightly dished. Back.—Great length; level and fairly broad across. Breast.—Full and nicely rounded. Wings.—Powerful; carried close to the body.

Tail.—Medium; gentle rise from back, increasing apparent length of bird; the drake having the usual curled feathers. Abdomen well developed.

Legs and Feet.—Medium, set wide apart. Feet straight and webbed.

Eyes.—Large and prominent. Keen and alert appearance.
Weight.—Ducks $4\frac{1}{2}$ to 6 lbs., Drakes $5\frac{1}{2}$ to 7 lbs.
Colour. Bill.—Pale yellow to deep orange. Head and Neck.—White, surmounted by a black cap covering the whole of the crown of the head to the top of the eyes.

Breast.—White. Back.—Solid black from the points of juncture of wing bows and body in a straight line across the back of the tail, and extending over the wings, giving the

BLACK AND WHITE MAGPIES
Probably the most popular variety of this breed of duck.
Their even markings make them great favourites

effect, when looking at the duck from behind, of a heart-shaped black mantle. The point of juncture of black and white should be sharp and clearly defined.

WINGS.—Primaries, white. Secondaries, white. The black upper part should show a clearly defined curve.

TAIL.—Black. THIGHS and RUMP.—White. EYES.—Dark grey or dark brown.

LEGS and FEET.—Orange. Black on legs and feet a slight defect.

SCALE OF POINTS

Shape, Carriage and Symmetry.	35
Cap, Head and Eyes	20
Colour (Body)	25
Tail	10
Legs and Feet	10

Serious Defects.—Wry tail; crooked back; slipped or twisted wings; feet that are not webbed; excessive weight; excessive coarseness.

THE CAYUGA

The Cayuga is both ornamental and useful, and a bird of good size, often weighing from 6 to 8 lbs., depending on sex. It is said that the Cayuga gets its name from a lake in America, that the first birds were imported to this country in the year 1871, and that these early specimens were not very brilliant in colour.

Breeders in England gradually improved them in size, type and colour and they are now very beautiful, especially the drakes, with their brilliant beetle-green feathering.

Great care must be taken in the mating up of breeding stock to avoid any purple tinge. Choose birds sound in colour, free from white feathers and with good primary and secondary wing feathers. Choose also birds of good body size combined with plenty of width, length and depth of body. But remember that a brilliantly coloured, moderate sized Cayuga will always beat a great big one which is dull in colour.

The flesh of this breed is white but unfortunately they seldom pluck out clean owing to the small black stubs, and

DUCKS

unless they are killed when in first full feather and showing hardly any stubs, they are hardly a commercial proposition.

If kept on and killed when in full adult plumage they pluck out a very good colour and show no dark stubs. They are docile birds, hardy, fair layers, and easy to rear.

Cayugas will successfully rear their own progeny, as does the wild duck, provided they are kept in a suitable situation with plenty of natural foods, and that vermin is killed; where necessary, supplementary foods should be supplied while the ducklings are in the young stage.

BLACK CAYUGA DRAKE

Although in few hands this breed carries valuable egg-laying and table qualities. It is also suited to crossing

The Cayuga is inclined to show white feathering after the first year; that is to say, perfectly coloured specimens will often moult and grow white feathers. This is not a sign of impurity of breed, but would seem to be caused by a lack of pigment, and often the white feathers increase at each yearly moult. An average useful mating is a drake and from four to even six ducks.

General Characteristics. HEAD.—Large. BILL.—Long, wide and flat, well set in a straight line from the top to the eye. EYES.—Full. NECK.—Long and tapering, and with a graceful curve.

BODY.—Long, broad and deep; prominent breast. KEEL.—Well forward and forming a straight underline from stem to stern. TAIL.—Carried well out and closely folded, the drake having two or three curled feathers in the centre.

LEGS.—Large and strong-boned, placed midway in the body, giving the bird a carriage similar to that of the Rouen. TOES.—Straight, connected by web.

CARRIAGE.—Lively, clear of ground from breast to stern.

PLUMAGE.—Bright and glossy.

WEIGHT.—Drake, 8 lbs.; Duck, 7 lbs.

COLOUR. BILL.—Slate-black, with dense black saddle in the centre, but not touching the sides, nor coming within an inch of the end, the bean black. EYES.—Black. LEGS and FEET.—Dull orange-brown.

PLUMAGE.—Very lustrous black, free from purple or white; the whole of the back and upper parts of wings, the breast and under parts of body deep black, the wings naturally more lustrous than the rest of the body. A brown or purple tinge is objectionable, although not a disqualification.

SCALE OF POINTS

Type	30
Size	20
Colour	15
Head and Bill	10
Condition	10
Neck	5
Tail	5
Legs and Feet	5

Serious Defects.—Red or white feathers, orange-coloured bill, dished bill; any deformity.

THE INDIAN RUNNER

Both from a fancier's point of view and from that of the utility man the Indian Runner is ideal, although it is not easy to breed a really outstanding Runner, perfect in colour and type.

The Runner is without a doubt the Game Fowl of the duck world—with its clean cut outline, short hard feathering, elegant carriage and graceful movements.

On the other hand, when bred for eggs, not forgetting type, it is certainly the Leghorn of the duck family—a really wonderful layer.

It is said that quite ninety years ago, Runners were brought by a sea captain into a port in Cumberland as a gift for his farmer friends; they were believed to come from Malaya and China. The captain's friends bred from them and also used them for crossing with other ducks. It was found that they were great layers and a big demand sprang up for these "Penguin Drakes," so called because of their upright carriage. Unfortunately this led to trouble, for any old drake or duck with a slightly upright carriage was sold as a Runner.

Evidently these ducks were then easily procurable abroad. A naturalist who saw them in their natural haunts as far back as the year 1856, wrote:—" The ducks are a peculiar breed, which have very flat, long bodies, and walk erect almost like penguins. They are of a pale reddish ash colour and are in large flocks . . . but are more generally known as Penguin Ducks."

From this it would seem that the Runner was quite common in its native haunt. Such is not so now, but quite a few importations have been made since the original one. Some were imported in 1924, but they were most difficult to procure. In 1909, Mr. Walton made importations direct and the birds were a vast improvement on the so-called Runners in England.

The first time Indian Runners were exhibited was at Kendal

in 1897. They were shown in pairs and the judge had 21 pairs on which to pass his opinion. No Whites were exhibited; it is fairly certain that they were non-existent at that date.

One thing certain is that the Indian Runner is not a breed made by the fancier; improved if you wish, but not a manufactured breed. It is still possible to import Fawn Runners which can win in the very hottest company.

The best way to understand the points of a Runner is to breed them. It is not easy to breed a real " top notcher " in any of the colours, and we have Fawns, Fawn and Whites, Whites, Blacks and Chocolates.

The Blacks were bred by mating a brilliantly coloured Black East Indian duck with a Fawn drake and vice versa. The mating gives Chocolates as well as Blacks; also of course, many poorly coloured birds of nondescript colours—muddy Fawns and poorly coloured Blacks. One also loses type, but this is put right by careful in-breeding.

The beginner always thinks of the Runner as a very long, slim bird and such are easy to breed. One must, however, breed a bird of medium size, not too big and coarse but, on the other hand, not a weedy little bird with no stamina. Start to breed a medium-sized bird with short, crisp, glossy feathering, clean cut in outline, compact in stern, and with nice length of body and a natural carriage of at least 60-70 degrees. By this is meant carriage maintained when in a pen in the open and on the alert, not in a show pen.

In a good Runner the body must be perfectly round—not flat on the back or chest. Get either of these faults in a bird and it is certainly not a good Runner either from an exhibition or a utility point of view, for such a bird will lack lung and heart space.

The eye must be bright and prominent and placed high in the skull, as near to the top of the head as possible. The skull must be lean and rather flat on top. A coarse, round skull and the eye low spoils the appearance in that it gives what we term a " clock faced " Runner.

The bill should be of medium length, strong, and fit imperceptibly into the skull; if you get a weak bill which is concave on the top line you do away with the wedge-shaped skull and bill and get what is termed a "dished bill" a very bad fault in a Runner. On the other hand, if the bill is too strong and inclined to be convex on the top line you have what is called a "Roman nosed" bill.

The great essential is to breed for the happy medium; in other words, one wants a clean line from the top of the bill to a point over and behind the eye.

The Runner is a bird of movement and without good sound legs and feet it cannot move with what is termed a perfect "outrun"; going away or coming towards you the bird must not "waddle" as does an ordinary duck, it must go with a perfect outrun, head up, stern down, upright carriage, perfect balance, and without the slightest trace of a waddle.

The legs of a true Runner are set on to the body much further back than in the other breeds. I like legs of nice length with good sized feet and strong thighs. Given these and you get movement and a real Indian Runner.

As a layer I consider the Fawn and White to be about the best, closely followed by the White. The former usually beat the Whites because they are not quite so temperamental and nervous—but much depends on strain.

On free range the Runner takes a lot of beating, ranging far and wide and picking up much of its living at certain seasons of the year in the form of natural foods. On the other hand, they seem to do very well without swimming water and are quite fertile on land.

The Standard of Perfection laid down by the Indian Runner Duck Club is as follows :—

General Characteristics.—The Indian Runner, as compared with the larger domesticated varieties, is a small, hard-feathered duck of a very upright carriage, and active habits.

Its body appears elongated and somewhat cylindrical, the legs being set on very far back.

NECK.—Fine, long and graceful, and when the bird is on

DUCKS

White Indian Runner Duck

Fawn Indian Runner Duck

the move or standing at attention is carried almost in a line with the body, the head being carried high and slightly forward.

BODY.—Slim, lengthy and rounded. From the shoulder points to the hip joints it should be nearly a cylinder. It is a little flattened across the shoulders, however. At the lower extremity the body sweeps round gradually to the tail, which is neat and compact and carried almost in the body line or horizontally, but should not be elevated or tilted upwards. The position or carriage of the tail varies with the attitude of the duck, but habitually upturned sterns and tails are objectionable.

The stern appears short compared with other breeds. The prominence of the abdomen and stern varies in ducks according to the season and the age of the bird, being fuller when in lay, but moderation in this point is to be observed. The large pendulous abdomen, often accompanied by long stern, and the other extreme the " cut away " abdomen and stern are to be avoided.

WINGS.—Small in proportion to the size of the bird, are carried tightly packed to the body and well tucked up. The tips of the long flights of opposite wings cross each other over the rump of the bird, more particularly when standing at attention. At the upper extremity the body contracts to form a funnel-shaped process, which again gradually and imperceptibly, without obvious junction, merges into the neck proper, which is long and slender until it again expands slightly to be fitted neatly into the head.

The lower or thickest portion of this funnel-shaped process or neck expansion should be reckoned as part of the body, and the distance from the top of the skull to where the neck proper joins the thick part of the funnel should be about one-third of the total length of the bird.

The thinnest part of the neck is approximately that part where, in drakes, the dark bronze portion of the head and upper neck joins the lower or fawn portion of the neck proper. The neck should be neatly fitted to the head so as to maintain a clean, racy appearance. The muscular part of the neck should be well marked, rounded, and stand out from the windpipe and gullet, the extreme hardness of feather helping to accentuate this. The long axis of the body should pass through the centre of the funnel-shaped process (or neck-expansion), but as the head is carried high and slightly forward the greater part of the head will lie in front of this axis.

HEAD.—With the bill, wedge-shaped, lean and racy looking. The skull is flat on the top and the orbit, or eye-socket, is set extremely high so that its upper margin seems almost to project above the line of the skull.

EYE.—Full, bright, and alert.

BILL.—At its base should be fitted imperceptibly into the skull.

LEGS.—Set far back on the body so as to allow of an upright carriage. The thighs are longer than in most ducks and are strong and muscular; the shanks are short and the feet neat and supple.

TOTAL LENGTH.—Drakes, 26 to 32 inches. Ducks, 24 to 28 inches.

WEIGHT.—Drakes, $3\frac{1}{2}$ to 5 lbs. Ducks, 3 to $4\frac{1}{2}$ lbs.

COLOUR. THE FAWN.—In Fawn ducks the general coloration should appear an almost uniform warm ginger-fawn throughout, with no marked variation in shade in the different parts of the body. The wing-bars are of the same colour but a few shades darker. The body exhibits a slightly mottled or speckled appearance. The bill is black, feet and shanks black or occasionally dark tan. The iris of the eye a golden brown colour.

When taken in hand the head and neck, also the lower part of the chest and abdomen, may appear a shade lighter than the rest of the body. Each small feather of the head and neck is lined with a fine line of dark reddish-brown, giving the head and neck a ticked appearance. The lower part of the neck and the neck-expansion are a shade warmer in colour, each feather pencilled with warm reddish-brown. In young ducks which have just completed the first adult moult there is often a rosy tinge on the lower neck-expansion, upper part of breast, and shoulders, but this soon fades away. The scapulars (or long pointed feathers on each side of the back covering the roots of the wings) a rich ginger-fawn, just a shade darker than the shoulder and back, with a well-marked red-brown pencilling.

The body of the wing or bow is a shade lighter than the scapulars but darkening towards the wing-bars, the feathers pencilled as before.

Some ducks have a cream or light coloured narrow band in the wing-bar owing to the upper part of each feather being of a lighter or almost cream shade edged or laced again with the normal dark shade. The secondary flights are of a warm red-brown.

The primary flights are usually a shade lighter than the secondaries. The shade of colour darkens and is richer on the back and rump of the bird, the pencilling also being richer and more marked, but the ground colour becomes lighter and warmer towards the tail. The feathers of the tail are lighter in shade, each feather being pencilled.

The belly-colour is lighter than that of the upper parts of the body, being about the same shade of fawn as the head and neck, becoming a shade darker on the tail cushion, all feathers being pencilled as before.

FAWN DRAKES IN FULL PLUMAGE.—The head and upper part of neck are of a dark bronze colour with metallic sheen, which may show a faint green tinge. This dark portion should meet the colour of the lower part of the neck with a clean cut or the lower colour merge into it imperceptibly. The lower neck and neck-expansion are of a rich brown-red which is continued on to the breast and over the top of the shoulders and upwards to where it joins the head and neck colour, merging gradually on the back and breast into the body colour. Sometimes this brown-red or "claret" is absent, the French-grey of the lower chest and abdomen extending right up to the bronze of the upper neck. The lower chest, flanks and abdomen are of a French-grey, which is made up of a very minute and dense peppering of dark brown, or almost black, dots on a nearly white ground, giving a general grey effect without any show of white. This grey extends beyond the vent until it meets the dark or almost black feathers of the cushion under the tail.

The scapulars are reddish-brown, peppered. The back and rump deep brown, almost black. The fan-feathers of the tail are dark brown, almost black, as is also the tail curl.

The body of the wing, or bow, is fawn, not pencilled. The bar is fawn corresponding with the wing-coverts in the lower part, the upper part is darker brown in colour, corresponding with the secondary flights, which are of a black-brown with slight metallic lustre. The primary flights are brown, fairly dark in shade. The iris of the eye is often darker than in ducks. The bills vary from pure black to olive-green mottled with black and black bean. The legs may be black or dark tan mottled with black, as also the feet and webs.

When in eclipse or duck plumage the drake more closely approaches the female in colour. All the dominant colours fade, but the head and neck are darker than in the female. The body colour becomes a dirty fawn or ash colour, with perhaps some rustiness on the breast.

FAULTS IN COLOURING OF FAWNS.—White anywhere; eye brows or eye stripes; light or cream wings (bows, coverts and flights) in the duck; blue or green wing-bars; orange or yellow bills, feet, or legs.

THE FAWN-AND-WHITE.—The head should be adorned with a fawn cap and cheek markings, nearly the same colour as that of the fawn of the body in the case of ducks but of a dull, bronzy green shade in the drake. The cap is separated from the cheek markings by a projection from the white portion of the neck, extending up to and in most cases terminating in a narrow line more or less encircling the eye. The cap should come round the back of the skull with a clean sweep. There should be no " tails " to it. The cheek markings should not extend on to the neck.

DUCKS

The bill should be divided from the head markings by a narrow prolongation of the neck-white, from one-eighth to one-quarter of an inch in width extending or projecting from the white underneath the chin.

When the bird is young the bill is light orange-yellow, but soon green spots show themselves and gradually unite extending over the whole mandible so that it becomes entirely, or almost entirely, of a dull cucumber shade in the duck, and a greenish-yellow in the drake. This is what should be bred for, though a darker shade is permissible.

The neck should be pure white to about where the funnel-shaped expansion begins, and should meet the body-fawn with a clean cut.

The fawn should be uniform throughout the body—a soft warm or ginger fawn—the shade, however, will depend to some extent upon the amount of fading and bleaching caused by exposure and sunlight, which rapidly destroys the colour.

Colour uniform to the skin, the undercolour not of a different kind. This description applies to the fawn over the whole surface plumage except the rump and tail of the drake, including the under surface of the tail, which are of a similar hue to his head.

When closely examined the coloured body feathers of the drake show a soft warm ground colour slightly peppered with a rather warmer shade. As only the outer edge of the feather is visible the colour seems solid and more ruddy than in the duck.

The duck should have much the same shade of fawn as the whole Fawn duck. The fawn and the white should meet on the breast with an even cut about half-way between the point of the breast bone and the legs.

The base of the neck, upper part of the wings, back and tail should be as nearly as possible the same colour as the fawn of the breast, and from the fawn of the back an irregular branch is thrown off on either side and extends downwards on the thighs to, or nearly to, the hough. The white of the breast extends downwards between the legs to beyond the vent and may overlap the thighs in part.

If the bird is coloured between the ribs and thigh it is termed "foul flanked." The primary and secondary flights should be pure white, as also the lower part of the wing-bow. The legs and feet should be orange-red.

THE WHITE.—Should be a pure white. The iris of the eye is blue and the bill, legs and feet orange-yellow.

Stains or black streaks on the bills or mottled legs and webs are minor faults, for which a point is deducted.

THE BLACK.—Black Runners should be a solid black throughout, and have a metallic lustre like the Black East Indian. Bill, legs and web are also black or dark tan.

THE CHOCOLATE.—This is an offshoot from the Blacks and should be a rich even chocolate throughout. The drakes on assuming adult plumage become darker than the ducks, but the ground-work is the same. The bill, as also the legs and webs, are black.

SCALE OF STANDARD POINTS

Head, eyes, bill and neck (exclusive of lower neck expansion)	20
Body, shape, and general appearance of (including lower part of neck, legs and feet)	35
Carriage and action	30
Colour and condition	15

Faults.—Above the Standard weights and measurements, short squat bodies or oval-shaped bodies, long stern, wry tail, slipped wing or any deformity, domed skull with central position of eye, dished and weak bills, Roman nose, under-curved bills, thick necks and short necks, swan or curved necks, the neck-expansion set on too far back on the body causing a chesty appearance with a hollow behind, flattened backs or bodies, legs set on too far forward causing poor carriage; waddling or rolling action, a natural carriage in any duck below the minimum 40 degs.

THE KHAKI CAMPBELL

The Standard laid down by the Khaki Campbell Club

gives every detail and feature of this wonderful breed. They are extraordinary layers, and if of correctly bred strains they will give almost unbelievable results in number and quality of eggs.

The Club Standard is as follows :—

General Characteristics.—The main characteristic of the Khaki Campbell duck is its wonderful power of egg production.

Other characteristics are : The pearly white colour of the eggs, which should never be green. Its extreme hardiness, caused originally by the use of wild duck blood and since preserved, partly by the avoidance of in-breeding, and partly by the open-air methods adopted by breeders.

The delicate flavour and quality of the meat, brought about mainly by the presence of wild duck blood. The suitability of the type and body formation for sound health and egg production. The serviceable and useful colour of the plumage which does not easily show discoloration through mud and rain.

Mrs. Campbell, the originator of the breed, wished to produce a useful, actively foraging, moderate-sized, good looking duck, capable of producing a large number of good sized pearly white eggs. She did not wish Khaki Campbell ducks to become an exhibition breed, mainly because breeding for the show pen often causes too close in-breeding, and birds are frequently kept in unnatural environments, two conditions which eventually always damage constitution and egg production.

This original aim and policy were steadfastly followed by most of the earlier breeders and have been strongly upheld by the Khaki Campbell Duck Club.

The original simple colour has been amplified in order that breeders may learn more about the breed and appreciate it better. The amplification has not been made with the idea of causing the Khaki Campbell to become a popular show bird, which it never was and never ought to become.

In type the Khaki Campbell is distinctive. Taking the

general outline in both sexes, the head is carried high, with shoulders higher than saddle, and with the back showing a gentle slant from shoulders to saddle. The neck is of medium length and almost erect, yet gracefully moulded. The chest is well rounded and prominent and, with the underline from chest and stern being somewhat rounded, the whole body appears compact or slightly compressed, while retaining depth throughout, especially from shoulders to chest and from middle of back through to thighs. The legs are not too far back, so that good abdominal capacity is noted in the rear without excessive sagginess in the ducks.

HEAD POINTS.—Bill proportionate, medium in length, depth and width, well set in a straight line with top of skull. EYES.—Full, bold and bright, showing alertness and expression, high up in skull and prominent. HEAD.—Refined in jaw and skull. FACE.—Smooth and full. NECK.—Slender, almost erect, medium length and refined.

BODY.—Deep, wide and compact. FRONT.—Broad and well rounded. BACK.—Wide flat and medium in length, gently sloping with shoulders higher than saddle. ABDOMEN.—Well developed at rear of legs but not sagging. UNDERLINE of BREAST and STERN.—Well rounded to permit depth through body at shoulders and from back to thighs.

CARRIAGE.—Alert, slightly upright and symmetrical; head and shoulders carried proudly; legs medium in length and well apart to allow of good abdominal development, not too far back. PLUMAGE.—Tight and silky.

WINGS.—Carried close and rather high. TAIL.—Short and small, rising slightly with usual curled feathers in drake's tail.

While aiming at good body size emphasis must be placed upon quality or refinement in general. By this is meant neat bone, sleeky, silky feathers, smooth textured face, refined head points, etc., with absence of coarseness, sluggishness, etc.

WEIGHT.—$4\frac{1}{2}$ lbs. for birds in laying condition in their prime.

COLOUR. THE DRAKE.—Head, neck, stern and wing-bar bronze; brown shade preferred to green bronze; rest of body

UTILITY KHAKI CAMPBELLS

Contrast these with the exhibition types of the same breed reproduced in the chapter on "Showing"

an even shade of warm khaki, legs and feet dark orange, bill green, the darker the better.

THE DUCK.—Khaki all over, ground colour as even as possible ; back and wings laced with lighter shade of khaki ; lighter feathers in wing-bar are permissible ; bill greenish black ; legs and feet as near body colour as possible ; head plain khaki ; streak from eyes considered a fault.

SCALE OF POINTS

Type (shape and carriage)	25
Colour	25
Quality or refinement	15
Head points	10
Size and symmetry	10
Condition	10
Legs and feet	5

Serious Defects. Yellow bill ; white bib ; any deformity ; green eggs.

THE WHITE CAMPBELL

The White Campbell is becoming very well known, and it deserves to be, for it is certainly a most useful variety—a white sport from the Khaki Campbell.

Those strains which are pure and of genuine full Campbell blood are excellent but, unfortunately, not all breeders have an eye for type and in many cases a rush to produce a White Campbell in a hurry to meet a big demand has meant foreign white blood being used.

Often small Aylesbury types of duck have been mated with White Campbell drakes. In other cases the Aylesbury blood is used with the idea of improving size, so that the resulting progeny will be a better table duckling.

This, surely, is a big mistake, as the idea of the White Campbell was to produce a pure white bird with all the good laying properties and type of the Khaki but with the advantages of white feathers and light-coloured skin and flesh. This is so that young drakes and old stock when scrapped will realise

DUCKS

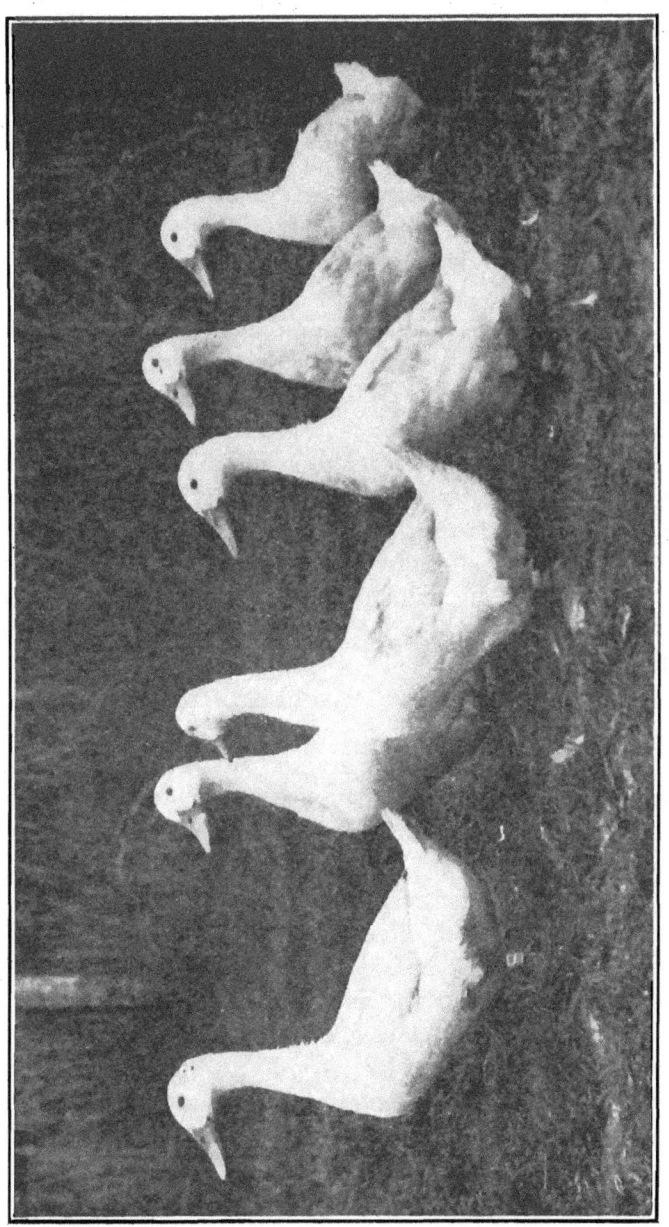

WHITE CAMPBELLS
The latest variety to be added to this family. In the opinion of many the White has a great future facing it

a better price per head; also, the white feathers are much in demand.

The White Campbell should be kept pure and of the same type and size as the Khaki. Do this and you have the White Leghorn of the duck world—vigorous, hardy, fertile on land, highly fecund and with great winter laying properties. Why spoil it by bringing in Aylesbury or Pekin blood?

The White Campbell might certainly be mated with smallish, active Aylesbury or White Pekin drakes early in the year to produce table ducklings, but the progeny should be killed and certainly not sold as White Campbells. The table breed drakes should be withdrawn and pure White Campbells introduced in time to give April and May pure-bred ducklings as future layers and stock.

The White Campbell has come to stay. The great essential is to purchase only the perfectly true type when starting and to breed up your own stock.

The White Campbell Standard is similar to that for the Khaki except for colour, which is as follows: BILL, LEGS and FEET.—Orange; EYE.—Grey-blue; PLUMAGE.—Pure white throughout.

Serious Defects. Excessive weight or coarseness; flesh coloured bill; any deformity.

The Campbell drake is light and active and may safely be given six or seven ducks. Of course, if an old drake, judgment must be used and the ducks reduced to a suitable number.

THE MUSCOVY

It is to be very much questioned if the Muscovy really is a duck! It is more like a goose in more ways than one; for instance, it is a grazer and eats grass in the same way as a goose. This is really an asset and ensures that the Muscovy is not costly to keep. They are wonderful fliers, very strong and powerful in flight, yet very tame, so tame as to become rather a nuisance, and so friendly that they have the habit of getting in the way of one's feet at times.

Unlike the drake of other breeds, the Muscovy male has no

DUCKS

curl feathers in his tail. This also points rather to the goose family. Again, Muscovies lay much as the goose does. They will produce a nest of eggs and then sit, or, if the eggs are withdrawn, leaving two or three of the freshest so that the duck continues to lay in the nest, she may lay 20 to 25 eggs and then go broody; alternatively, if the nest is broken up when she does commence to sit, she will rest for a time and then lay again.

Unlike other ducks the incubation period is 36 days; also, the ducklings are not in first full feather until 16 weeks

BREEDING PEN OF MUSCOVIES
Doubt, says Mr. Appleyard, still exists as to whether the Muscovy should be classed as a duck

of age, whereas the ordinary duckling is in full feather at 12 weeks.

Another point, when the Muscovy is mated with domesticated ducks, say Aylesbury or Pekin, the progeny are mules, birds which are sterile and will not breed. This again would seem to indicate that the Muscovy is not a true duck.

Many writers say that the Muscovy is most pugnacious and troublesome with other birds, but with this statement I cannot agree. On my own farm many of them are at liberty, not wing-clipped or pinioned, and they go where they like and have never been any trouble to or harmed any other bird.

It must, of course, be clearly understood that they have always been mated, not just an adult drake or two without any females; such a state of affairs might cause trouble.

Muscovies are quite safe to run with any other waterfowl on the farm if mated, and they will not then inter-mate with other ducks. The ducks are wonderful sitters and mothers, and this sitting feature can be made good use of.

The ducks always choose a good place for their nest, safely sheltered and secluded, and they will sit week after week. They will take 12 to 15 eggs, or even more, and so are most useful as broodies, removing the duck eggs about the twenty-third or twenty-fourth day and giving more fresh eggs; the other eggs are finished in the incubator or under hens.

Another noteworthy point is that the drakes are very big, often weighing 10 to 12 lbs., whereas the duck is very small, about $4\frac{1}{2}$ to 5 or 6 lbs.

There are many different colours and markings to be had, viz.: Black and White, Blue and White, Dun and White, pure White, Black and Bronze. Many are now bred with clearly defined markings, such as Magpie markings.

The Muscovy is the common duck of Central and South America and the West Indies. It is a native of Brazil. In Australia it is very popular, and is used just as we use the Aylesbury or Pekin in England; in fact in Australia it is preferred to the White Aylesbury or the Pekin, and is the table duckling of that country.

In appearance it is a peculiar, yet striking looking bird, as its face (or cheeks) has no feathering and shows a rich red or scarlet skin or fleshy space around the eyes, and the base of the bill is surrounded by scarlet.

The drake is, of course, much more striking in appearance, being big and with his head rather large in proportion to his body. On the head are long feathers, which he can raise or depress at will. This he does when mating, or if alarmed, excited or annoyed. The drake is voiceless except for a very low harsh hissing sound, and this he only makes use of if scared or annoyed.

The duck can quack but hardly ever does so, and even when she does, it is only a very low, quiet sound. When with her ducklings, or if broody, she makes a peculiar whistling noise.

It is a grand sight to see a Muscovy duck with 12 to 14 small, hardy ducklings; generally, she makes a huge success in rearing them. A drake will mate with just one or up to eight ducks. Personally I usually allow three or four ducks to a drake, and the eggs are nearly always 100 per cent. fertile and hatch 100 per cent. Young Muscovy is delicious eating, of excellent flavour and without much fat; it is rather "gamey" and there is plenty of breast meat.

In conclusion, a word of warning on handling this breed. Muscovies are armed with very long and sharp claws, like talons, and are quite capable of opening up your wrist or hand. If you must handle, do so by grasping the bird firmly by the wings, where they join the body, and by the neck, but keep out of the way of their claws.

There is as yet no club or Standard for this breed, although they are exhibited at most shows where A.O.V. duck classes are scheduled. At the London Dairy Show there are classes for both sexes.

THE CRESTED DUCK

The Crested is now quite an uncommon breed, although from time to time one sees a good specimen at the shows. They are rather like an Orpington duck in size and type, the chief feature, of course, being the crest. They may be any colour, white, blue or white splashed with colour. The crest should be round, large, and carried on the top of the skull; in appearance rather like a powder puff.

It is certainly an interesting breed for those who wish

to keep a few ducks as a hobby, and for general home use plus the added interest derived from the difficulty in breeding a good shaped bird with a really good, round, shapely, well-placed crest. This is not easy.

In breeding Crested ducks only a percentage come with crests, but when hatched the plain headed ducklings can at once be distinguished.

General Characteristics. HEAD.—Long and straight. CREST.—Globular, large, set evenly on the skull. BILL.—Long and broad. EYES.—Large and bright.

WHITE CRESTED DUCKS
As stated by the author there are a number of other colours in this breed

BODY.—Long, broad, and fairly deep; full, round breast; long broad back; strong wings, carried closely; short tail, similar to that of the Aylesbury.

LEGS.—Short and strong. TOES.—Straight, connected by web.

DUCKS

CARRIAGE.—Somewhat erect.
WEIGHT.—Drake 7 lbs., duck 6 lbs.

SCALE OF POINTS

Crest	25
Type	20
Size	15
Head and Bill	10
Colour	10
Condition	10
Neck	5
Legs and Feet	5

Serious defects.—Slipped wings; any deformity.

THE BLACK EAST INDIAN

A small beetle-green coloured bird, hardy and a free breeder, but only a fair layer. The smaller in size, and the richer in lustrous green sheen, the better. An ideal breed for the hobbyist, as it gives both beauty and hardiness and surplus birds make delicious eating; dark-fleshed but with a flavour all its own; very plump-breasted and light in bone.

East Indians will lay, sit, and hatch and rear their young in suitable surroundings. Alternatively the eggs may be taken away and placed under a light hen or in an incubator, in which they hatch well.

A drake will mate happily with one duck or with up to four or five. Personally I prefer one drake with two or three ducks. They do little harm in a large garden, and will clear up every slug, snail, worm, etc., they can find.

If you have a large garden with lawns and shrubberies, a trio, or a drake and three or four ducks, will live wild; there is no more delightful sight than a Black East Indian duck with a brood of ten or twelve tiny beady-eyed, black ducklings darting about catching flies. It is best to pinion them when hatched, as when in feather they are excellent fliers.

General Characteristics. HEAD.—Neat and round with high skull. BILL.—Short and fairly broad, well set on a straight line from tip of the eye. EYES.—Full. NECK.—Short.

BODY.—Short, broad back, round and prominent breast.
LEGS.—Medium length, placed midway in the body.
TOES.—Straight, connected by web.
CARRIAGE.—Lively, smart and symmetrical, clear of ground from breast to stern.
PLUMAGE.—Bright and glossy and very tight.
WEIGHT.—Drake 2 lbs., duck 1½ to 1¾ lbs.
COLOUR. BILL.—Slaty black, olive patches not objectionable. EYES.—Dark and full. LEGS and FEET.—As black as possible, becoming lighter, more or less orange, with age. PLUMAGE.—Very lustrous, intense beetle-green, free from purple or white feathers.

SCALE OF POINTS

Type	20
Size	20
Colour	30
Head, Neck and Bill	15
Legs and Feet	5
Condition	10

WHITE AND BROWN DECOYS

These are charming little ducks, often called White Call or Quack ducks. The White is fairly common but the Brown is rare; the reason may be that the Brown to the ordinary eye is much like the Mallard in colour.

The chief points in a good Decoy of either colour are size and type. The ideal Decoy is very small, very compact, clean cut, and with a really cobby body. The eyes are dark and alert, and set in the middle of the face.

The White must be white—pure Chinese white—free from sappy colour. The feathering must be short, glossy and crisp. When mating, pick the very smallest ones with broad, deep bodies—those with rounded skulls, short, thick bills of a bright orange-yellow, and legs and feet of the same colour.

The Brown should be an exact copy of the White except in colour. Avoid the thin-bodied, long, boat-shaped birds with flat skulls and thin, longish bills.

DUCKS

Usually, Decoys mate best as a pair or in trios. They are excellent parents and will sit, hatch and rear their own young most successfully in the open so long as it is not too early in the season. Generally speaking it is best to keep taking the eggs early in the season to be hatched out under a small hen or bantam.

CROSSBREEDS

As in all live stock, there are times when one can gain some advantage by crossing certain breeds.

In table breeds, when one wants to produce a large number of quick maturing ducklings, a cross between the Aylesbury and the Pekin is quite useful, for an out-cross generally gives stronger and easier-reared progeny—if, of course, both sides of the cross are strong and vigorous.

A TYPICAL UTILITY CROSS
Left: Aylesbury; Right: Pekin. These ducks, as outlined in the chapter on Crosses, make an admirable table mating

The Pekin drake is usually very active and fertile in the early season and if mated with Aylesbury ducks the fertility is generally very good; the progeny are rapid growers and mature early. Alternatively, the Aylesbury male can be crossed with Pekin ducks; as the latter are usually better layers than the Aylesbury it gives more eggs to sit.

We also have the Rouen-Aylesbury and Rouen-Pekin crosses—generally used by private owners who wish to produce ducklings for their own table—not too dark fleshed, and with a rather gamey flavour.

Another useful cross for those who keep layers such as Runners and Campbells is to use with them a medium sized, active Aylesbury or Pekin drake from large parents.

The Muscovy drake will cross with any good sized duck so long as he has no ducks of his own breed. The Muscovy duck is fertile with the ordinary drakes only if no male of her own species is around. The resulting progeny are mules and will never breed.

For my own personal family use the choicest duck ever is the progeny of, say, a Mallard drake and White Campbell ducks; this is only my personal opinion and the results tried out on friends gave great satisfaction; or you can use Khaki Campbell ducks but the progeny is darker in flesh colour.

The progeny from White Campbell females is beautiful to look upon, lovely shades of silver and golden down. To eat they are truly delicious. The idea is to rear them as good as wild—the ideal place is on a farm pond. Grain fields and much natural foods such as acorns, beech mast, etc.; then commence eating them in October and onwards, roasted complete with rich brown gravy, sage and onion stuffing, apple sauce, white potatoes and vegetables such as celery, mashed swedes or brussels sprouts. The ducks to be well cooked and very brown. Hot they are grand, cold they are wonderful!

Chapter IV

HOUSING ADULT STOCK

IN many districts it is absolutely necessary to house ducks. First, there is the question of the fox and large vermin.

Secondly, the birds may be on land which is very exposed or in a very cold district.

Let us examine this question of the best type of house to use. In the first place, light in a duck house is of no advantage, for the ducks do not use the house during the daytime; as a matter of fact lots of windows and light in a duck house are a great disadvantage. This may sound rather silly, but in practice it will be found only too true. On moonlight nights windows throw shadows on the floor and keep the ducks on the move: nervous members of the flock keep working about and disturbing the more restful birds. Or, it may be that ducks are really night feeders and the moonlight disturbs them, making them wish to be out and working for natural foods.

So I advise little light in the house and as much ventilation as possible without direct draughts. Next, we do not require lots of head room to work in, as we have only to clean the house and collect eggs, so that the expense of a high house is unnecessary.

Housed ducks must have a dry floor, otherwise the whole atmosphere quickly becomes hot and smelly, causing the birds to do a partial moult and go out of production.

It is often possible to make good duck houses from buildings already on the place, it being up to the owner to use his own ideas in converting such buildings.

Remember also to have a good wide door. Small, narrow entrances are fatal to ducks, for they generally go in and out of their house with a rush. Also remember that ducks are not acrobats!

HOUSES WITH WIRE FLOORS. From a practical point of view these are not a success with adult ducks; the birds dislike them and are unhappy and thus do not give good results. They may be used with bedding on the wire but even then they are difficult to clean out. Wire floors without bedding, of course, mean many broken eggs.

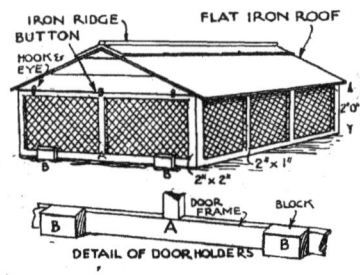

SLATTED FLOOR HOUSES OR ARKS. These again are not a success. It means bedding on the slats, and the house is even more difficult to clean out than one with bedding on a wire floor.

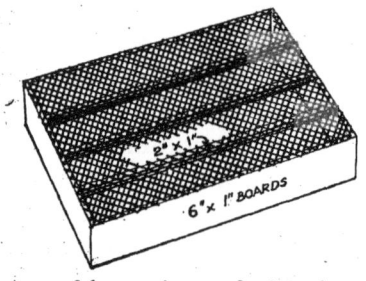

A useful vermin-proof night house for partly reared ducklings in hot weather. It is fitted with the wired platform seen in lower half of sketch

WOODEN FLOORS. This would seem to be the best and most satisfactory flooring for the average small house for adult ducks. When houses have wooden floors, it is an advantage to have the floor just off the ground, with a good air-space between the bottom and the ground.

CEMENT FLOORS. These are most useful, especially in larger and permanent buildings, as, when properly laid, they are rat-proof and can from time to time be given a really good wash and be well disinfected. If rats get working under a cement floor, they can be easily dealt with by "gassing" or by dropping pieces of carbide into the holes and following this up with a little water. All this must be done, of course, when the ducks are out of the house for the day. Another way is to pour some crude tar into the holes.

Cement floors can be made in the form of platforms on

DUCKS 63

which to stand small duck houses. In this case the platform must be half an inch smaller than the house, or water will run down the house on to the cement.

When working with cement floors, it is necessary to keep them well bedded with deep, dry litter. Cement is an "unnatural" and cold floor, and it means many cracked and broken eggs unless it is well littered.

NATURAL FLOORS. The success of this type of house

AN ECONOMICAL SHELTER HOUSE

This photograph shows a handy type of shelter which can be erected at low cost on the smaller duck farms. It is admirably suited to growing ducklings

depends on the nature of the ground. In many cases it may be a success, but one difficulty is rats. With an ordinary earth floor, these pests are difficult to keep within bounds, and if in large number the ducks are disturbed during the night and consequently their egg yields are poor.

Permanent and very good floors may be made from local materials, chalk being excellent for the job. I have used vast quantities of chalk, placed it in position, watered it, and then rammed and beaten it down. It sets like a rock and so long as you keep it well littered it is an ideal "natural" floor.

Another good floor may be made from engine ashes and tar mixed and well rammed. If it gets "sticky" while being put down, try a sprinkle of dry sand. Cut out all unnecessary expense in housing ducks; house them in decent comfort and then remember that it is careful management, good birds and proper feeding that give results.

DUCKS IN COMPOUNDS. I am including this under the chapter on housing, because the compound takes the place of a house. By compounding ducks is meant keeping a pen or flock in an open compound, a method that has proved successful over more than 15 years with many breeds and varieties.

It will work with any number of birds from a trio to a flock of 50 or more, but with laying or breeding stock the best number is 25 ducks and four or five drakes as a limit. The size of the compound depends, of course, on the number of birds you propose to place in it for the year, but they are best built about 8 by 10–12 feet. These accommodate up to 30 birds, yet will, of course, do for as few as a drake and five or six ducks.

Choose the driest and most secluded spot on which to erect the compound, preferably on high, dry ground. If there is natural shelter, such as a hedge, to give protection from the north and east winds, so much the better. Get the necessary poles or stakes, peg out the plan, drive in the stakes, and leave a good wide gateway. Use six feet netting of two-inch mesh. Peg the bottom selvedge of the netting to the ground; old large horse-shoes are useful for this purpose. Galvanized iron sheets may be used round the bottom of the pen if thought necessary, or if it is in an exposed position.

If you have an old ark or small house, without a floor, it will be used by the ducks to lay in, or along one side a rough frame-work may be run up, and a galvanized iron roof nailed on. Failing this, the ducks will lay behind a few sheets of

DUCKS

galvanized iron placed against the wire. A few old coops, boxes, or even small empty barrels will also be made use of by the ducks.

COMPOUNDS MADE FROM NATURAL MATERIALS. These can be made really cheaply, and, if anything, are better than the first-mentioned compound. On farms and in the country, ash, elm, nut and conifer poles may be made use of; faggots made from hedge-cuttings, bank trimmings, rushes, reeds, etc.

Choose your site, mark and peg out the plan, drive in the main poles, and across these fasten lighter poles, either by nails or wire, about three feet from the ground. On these are fastened the faggots. Next, dig out a trench nine inches deep and about 12 inches wide in which to place the butt ends

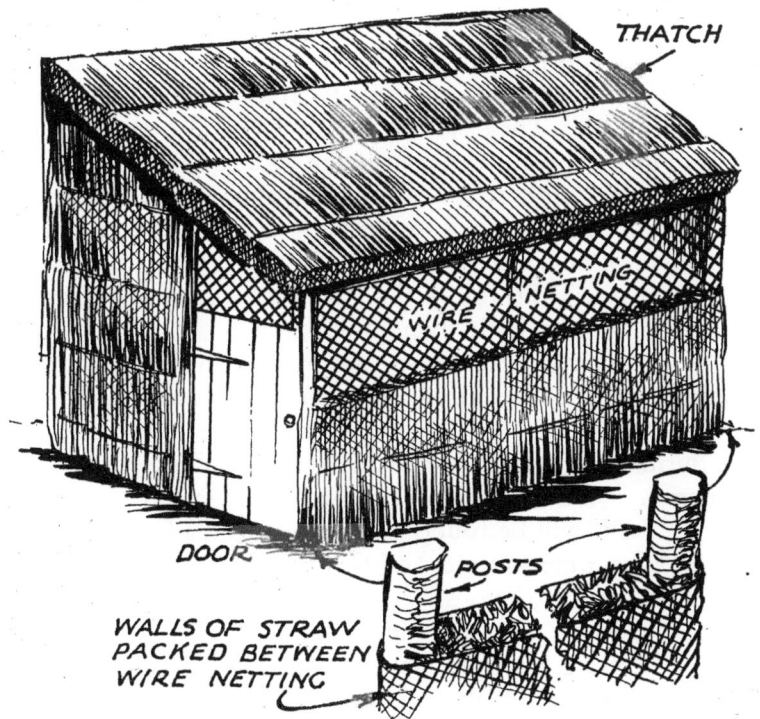

A home-made shelter such as is described in this chapter

of the faggots. Do not dig the trench at the gate entrance. The faggots should be made as tall as possible, from brushwood, rough grass, hedge trimmings, etc., or, if you can get reeds, they are next to perfection, and last for years.

Place the butt end of the faggots in the trench; work each faggot well into the other and wire them to the frame-work. Once you have them in position, take the earth from the trench and throw it on to the bottoms of the faggots, stamping and beating it down well.

Next, put in a gate, and if it is covered with galvanized iron so much the better, as it makes for strength and keeps out direct winds. Moreover, it completes the privacy of the compound.

If it is thought necessary, a length of two-feet netting of two-inch mesh may be run round the outside walls, but if the faggots have been properly made and well packed into position this will not be necessary. Anyone on the average general farm can get the material to erect these compounds, free of all cost. They can be improved if low shelters are arranged inside : again from natural poles and reeds or straw.

These compounds will answer admirably where the ducks have the use of a large pen or orchard, or absolute free range. If a number of flocks are run, several can be erected at different places, and it will be found that the ducks are easily trained to use and keep to their own compound.

When working with these compounds water and food containers will be required, and they must be filled each morning and evening. Cleaning is necessary only once per year, during August, when the ducks are moulting and out of production. This may sound most unsanitary, yet it is not really so, for it will be found necessary to add a fresh layer of bedding from time to time, and this forms a kind of drainage mattress.

The advantages of compounding ducks as compared with ordinary housing, or open pens, are manifold. There is no labour in frequent cleaning, those with little capital can keep many ducks without a lot of expense in houses, and there is no rapid depreciation which goes with such houses when used by waterfowl. The ducks live a very natural life at night without any risk of " fug " and sweating, which generally results in false moults. They seem happy, and are certainly less nervy than when under a roof. Best of all, they can have

A useful type of cheap shelter to use in a laying compound or as a shelter for strong ducklings. It will also serve for ducks to lay in when they run in open pen unhoused

water throughout the long nights. This alone is worth a lot when working with birds which love water and are night feeders. In a closed house, if water is given where there is a floor, it means a terrible mess.

When using a compound it is found that about 100 square feet is ample for 30 birds; even 40 without any harm. As to floor space per duck when houses are used, it is difficult to lay down any hard and fast rules. For instance, the house which would comfortably take 20 White Runners would be crowded if it had 20 large Aylesburys in it; and, if not too well ventilated it might not be a success with even six Runners in it!

A well-ventilated house six by four by four feet proves a success when housing up to 12 light breed birds (say two drakes and ten ducks) or with two drakes and eight ducks of such as utility Aylesburys or Pekins. From this one can work out how many birds can be put into any house making allowance for height and ventilation.

When working with laying or breeding ducks in a compound, the routine is briefly as follows : 7 a.m., replenish water in water tins and give a grain feed (see Chapter on "Feeding"). For preference place the grain in the water tins; 9 a.m., liberate the ducks, fasten the gate open securely and collect the eggs. In winter and early spring feed mash and water in tins one hour before dusk. Fasten the gate for the night after seeing that all the birds are in and in order.

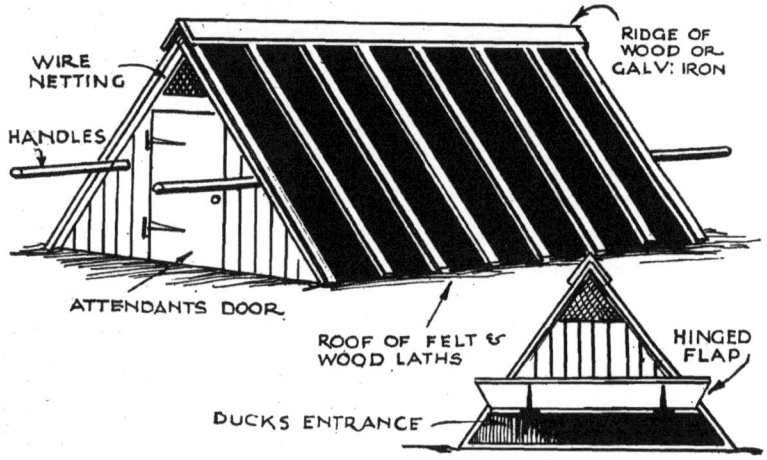

An Apex duck house. It has no floor and is easily movable. Can be used for either laying or breeding stock

During light evenings feed the mash at a fixed time, say 5 p.m., and leave the gate open. Just before dusk, go round and get the ducks into the compound, water them, and close up for the night. Usually the ducks "put themselves to bed" and the attendant has only to fill the water containers and close the gate.

CHAPTER V

FOODS AND FEEDING

FORTUNATELY, a duck is not a fanciful or dainty feeder. All it asks for is a regular diet of good plain food, carefully and properly mixed and fed at regular times. In many cases, and for quite six months in the year, the adult duck, if it has suitable land and water to work on, will pick up a big percentage of its living in the form of grass, water weeds, slugs, snails and insects. Also, when given the opportunity, it will pick up much useful food on corn stubbles. From an egg production point of view the secret of success with any duck is contentment, freedom from scares in the form of high winds, stray dogs, foxes, etc., suitable housing or compounds in sheltered positions, and good sensible foods—as much as they can reasonably eat.

The poorly fed duck cannot keep up production for any length of time. The well-fed bird will, day after day, week after week, give 95 per cent. to 100 per cent. production in the form of $2\frac{1}{2}$ to 3-ounce eggs.

FEEDING LAYING OR BREEDING STOCK.—Grain may be fed each morning, preferably in troughs, about 7 a.m. or, if wished, upon liberation at 9 a.m. In most cases, however, it is best to feed as early as possible in the morning, while the birds are in their house or compound. If you wait until 9 a.m., the usual time when laying ducks are liberated and eggs collected, it will usually be found that the ducks are so keen to get out on the range in search of natural foods that they will neglect the corn. On the other hand, when fed at, say, 7 a.m.—if possible in long water troughs, and the grain covered with water so that sparrows, rats, etc., cannot eat it—it gives the ducks an interest in life and does away with the strain which goes with awaiting liberation, plus the fact that they eat the corn and get nourishment and a good foundation for the day's work. Then, on liberation, the birds can go right away, on range, or into their pen as the case may be.

Wet mash should be fed one hour before dusk, again in troughs, and see that there is ample trough room, so that the mash is not knocked all over the place to be trampled into the ground. Arrange things so that all the ducks can feed comfortably at one time. Place water in the containers so that they can get a drink throughout the night and in the early morning. The duck is a waterfowl and in a natural state is a night feeder, and it helps production if water is allowed for the time the bird is in the house or compound.

It is unwise—almost cruel—to expect a duck to eat her supper of mash without water. Supply this in deep troughs or tins of such a shape that the ducks cannot take a bath or make a mess, yet can drink freely. It will pay to give this point every care and attention.

All mashes should be mixed to a crumbly state with cold water, not a sloppy or pasty mess. If anything, let them be on the dry side, yet not so dry that a lot gets wasted; the duck can only make a very poor effort at picking up dry mash or particles of meal with its broad, blunt bill.

An ample supply of grit should be provided in the form of limestone grit—a really good shell-forming material. Coarse water-washed sand and gravel are most beneficial, and are best kept in suitable containers in the house or compound.

From time to time when the birds are in very heavy production, limestone flour may be added to the mash as an aid to good-shelled eggs. The duck in production generally lays an egg per day, sometimes two in well under the 24 hours, and she must be provided with plenty of shell-forming material.

SUITABLE MASHES.—The following mashes have been tested and proved excellent: Weatings 60 per cent., broad bran 15, maize meal 15, white-fish meal 10. More maize meal may be used during very cold spells.

Another good mash is as follows: Superfine weatings 6 parts, bran 2, maize meal 1, white-fish meal 1.

It is essential that the very best white-fish meal is used, of 60 per cent. albuminoids. Sour or strong smelling fish meal is useless, does a lot of harm and taints the eggs badly. Ducks on free range should be fed on equal parts of wheat and maize as their morning feed.

Each evening, a feed of three to four ounces of wet mash per bird (by weight, before the mash is wetted) should be given. At no time can any hard and fast rule be given as to the exact amounts to feed; it depends on the season of the year and on what foods the birds can pick up. When in production you cannot overfeed so long as you give correctly mixed foods at the proper times. It is certainly best to feed what mash the birds will clear up each evening in 15 to 20 minutes.

It is easy in these days to get really good proprietary mashes, well mixed and made up from genuine materials. Many large firms now specialise in the mixing and sale of mashes for all purposes; these are offered at quite moderate prices and in sound clean bags, which have the advantage that no disease germs can come on to the farm from old sacks.

To my mind a good duck mash must contain a good percentage of roughage, especially if it is to be fed to birds confined in pens. In a wild state the duck devours huge quantities of roughage in the form of water weeds, etc.

FEEDING MOULTING DUCKS.—Generally, ducks moult during the months of July and August. At this time they are in rather a sorry plight and during the moult they must on no account be treated as purely food eaters and in the way, even if for the moment they may be out of profit and non-wage earners. By far the best way is to think of the future. With proper foods, care and attention it is possible to build them up and soon have them laying an abundance of autumn eggs and right through the winter. Help them to get their new plumage and to build up stamina for the future. The following mash is useful: Bran 4 parts, maize meal 2, superfine weatings 4, ground oats 2, white-fish meal 1, cod-liver oil 2 per cent.

Another useful tip is to boil some linseed. Place the linseed in a crock, boil gently for a long time, and keep adding water when necessary. You will finish with a jelly, which should be used twice or three times weekly in the mash; it is most helpful to the ducks in getting new feathers. A little flowers of sulphur once a week in the mash is useful as a blood purifier.

LAYERS CONFINED IN PENS.—The only difference here is that it is best to provide a small feed at midday to keep the birds happy and contented. This simply means giving about two ounces of the mash at midday and the remainder as usual in the evening. It is also wise to give chopped green food in the midday mash, to make up for the green food they miss on range.

One good scheme is to arrange matters so that the ducks have to get exercise by running from the water pots to the mash trough and vice versa—so have the food pots at one end of the pen and the water containers at the other. Also, the grain may be scattered about the pen so that the birds get some exercise in searching for it and picking it up.

FEEDING YOUNG STOCK FOR EGG PRODUCTION.—I have tried dozens of different methods and mashes in the rearing of ducklings, all except dry mash. One class of food I consider most helpful and essential to all mashes for successful rearing is really sound biscuit meal, an easily assimilated and digested food, prepared ready except for soaking or scalding, and an ideal " breaking up " material for the mash..

When preparing wet mash for ducklings, be sure to mix it well and into a really crumbly, friable state. Take the required biscuit meal in a bucket, scald with boiling water, and place it on the mixing board. Put some dry mash on the biscuit meal and mix until the whole is well blended and in a proper crumbly state; add more mash if too wet and sticky. For myself, I always like to add a little fine limestone grit and a sprinkle of coarse washed sand to the biscuit meal before adding the dry mash. In a natural state ducklings devour huge quantities of sand, small gravel, etc.

FIRST FIFTEEN DAYS.—For the first 15 days an excellent

DUCKS

feed is wet mash of best biscuit meal dried to a crumbly state with equal parts of Sussex ground oats and superfine weatings. Feed every two hours.

FROM SIXTEEN TO THIRTY DAYS.—Wet mash made up from the following : Biscuit meal dried off with home milled bran 2 parts, superfine weatings 3, Sussex ground oats 2, white-fish meal $\frac{1}{2}$. Gradually reduce the biscuit meal and make it wetter, so that it takes more mash to get it dried off to the desired crumbly state.

From the fourteenth day the ducklings will be getting strong and active. Feed early and late. Meal times may be reduced by placing more food before the ducks in suitable troughs at each feed—say four times a day.

THIRTY-FIRST TO FORTY-FIFTH DAY.—One part of maize meal can be added to the mash, and the biscuit meal reduced ; also, it is now quite safe to add some soaked crushed wheat. Get your miller to put a clean sweet sample of wheat through the oat crusher or roller ; crushed or rolled wheat is a grand food. The ducklings love it in their mash or even placed in a trough with water poured on it, as they get older.

GROWERS' MASH.—Our ducklings are now six to seven weeks old. We do not wish to force them too much, and so we feed a good growers' mash, plus some grain.

The following two have been proved all-round growing mashes : (*a*) Bran 3 parts, superfine weatings 3, maize meal 2, ground oats 1, white-fish meal 1 ; and (*b*) Bran 4 parts, superfine weatings 6, ground oats 2, soya bean meal $\frac{1}{2}$, white-fish meal $\frac{1}{2}$.

In all cases the owner must use his own judgment as to including other things in the menu. · As I have said before, I am a great believer in roughage for ducklings and ducks and I find it pays every time to feed a fairly bulky mash with roughage in it. Bran is in a way roughage, so also is chopped green food. This is most beneficial and when possible it should be given in plenty in the mash. Chopped lettuce, cabbage, lucerne, clover, even nettles and green grass, keep the food bill down and do much good from the health point of view.

Here we come to the question of grain for the ducklings. We often hear a duck-keeper say, " My six weeks old ducklings are always calling out and clamouring for food." Now to grow and make a success with waterfowl, they must rest and be contented. After watching many old and successful rearers in action, I am of the firm opinion that in most cases it pays to keep food nearly always before the duckling. It certainly does with table breeds which are to be ready for killing at 10 to 12 weeks of age. Certainly no food is left about for the birds of the air ! By keeping to a programme of a little drink, a little walk, a little food, a little rest, a little sleep and so on throughout their short life, we should have a fat duckling.

Naturally we do not wish to fatten and kill the females of the laying breeds. A good practice with the latter from, say, five to six weeks, is to commence adding some dry cracked maize and crushed wheat to the wet mash, especially at the last evening feed. It remains with them longer than mash and they are more contented throughout the night and during the day.

FEEDING TABLE DUCKLINGS.—The secret of success in feeding table ducklings, or any breed of duckling intended for the table, is to have them fat at from 9 to 12 weeks of age, i.e., as soon as they are sufficiently feathered to pluck out clean and reasonably free from stubs. This is immediately before or when just in full first feather. Miss this stage and they go into a moult, which means feeding them for weeks before they are again in feather.

A duckling comes into the world with down as a body covering, then it gradually grows feathers and usually is fully feathered when 12 weeks of age. From the killing point of view, one can commence when they are nine weeks old or just a little older, when the flight feathers in the wing are about three parts grown.

When the duckling is about 12 weeks old it immediately commences to moult the body feathers and to get a new lot. Some even moult three times before the autumn.

DUCKS

In all cases it is only the body feathers which are moulted, not the wing feathers. Thus it can easily be understood why the man who knows how to rear and feed them without a check and have them off his hands by the end of three months is the man who will make money from table ducks. Food costs money, and the ducklings are taking labour, space, etc., to run on until they are in second feather.

There is good money for those who can continually produce prime, good fleshed birds of 9 to 12 weeks old, but to do so means care and attention to detail, and having the right class of good laying, fertile Aylesburys or Pekins, or the first cross of these two breeds. The following mashes are good :—

First Feed to Seventh Day.—A little biscuit meal and the following mash : Superfine weatings, 40 parts, bran 20, barley meal 20, Sussex ground oats 20, dried milk 10, cod-liver oil 2 per cent.

Eight Days to Eight Weeks.—Weatings 30 parts, bran 15, barley meal 20, Sussex ground oats 20, meat and bone meal 10.

Last Two Weeks. Fattening Mash.—Superfine weatings 20 parts, barley meal or Sussex ground oats 60, meat and bone meal 15.

Proper foods play a great part in success with ducklings, yet food, to my idea, is not everything. Attention and care, well-mixed foods, cleanliness, the correct amount at the right times and the right class of stock all help to make the whole undertaking successful and profitable.

In concluding this chapter on feeding, it might be helpful to say a few words on the best tools, etc., with which to work. First, I would advise anyone who has a number of ducks to make or have made a mixing table. They cost little, last many years and are a boon to those who have wet mash to mix in any quantity, saving both labour and waste.

Next we require a light, well-worn shovel with which to mix the mash. Also a number of light buckets and, in the case of a good number of birds, a galvanized iron barrow with a good wide wheel. In some cases this can be filled direct from the table and taken to the birds. The table should be near the water supply, and if water can be led by a pipe to the table and through a tap, so much the better. Year in, year out, ducks and ducklings require mash, thus it pays to have your " kitchen " in good order.

Mash mixing is made easier by using an appliance of this type. The author describes its advantages.

Another useful gadget is a proper food carrier made from a pair of wheels and with a wooden frame in which are suitably sized holes, each to take a bucket. Such a carrier is very useful for certain sections on the farm, e.g., the small breeding pens. Each can have its ration in a special bucket filled at the food store.

FOOD TROUGHS

In certain circumstances it is safest and most economical to give all foods in suitable sized and shaped troughs. There are, or course, exceptions to this rule, as for instance, the ducklings' first few feeds, which should be fed on the ground level on clean grass. A round tin-lid or piece of slate will form a small table on which to scatter the mash. Best of all is the lid from a round confectionery tin, about eight inches in diameter. This is easily rinsed clean and can be thrown away when finished with.

Another exception to the rule is when feeding grain to certain types of ducks on a pond or stream. If the water is suitable, about six to twelve inches deep, and with a clean bottom, grain may be thrown in. The birds like bibbing for it and by this method you are not feeding sparrows and rats! Make sure the birds are hungry and do not give too much, or it will lie on the bed of the pond and go bad.

Another method of feeding grain under water of some depth is to build a wooden table-top with three-inch sides and ends. Stakes are driven into the bed of the lake and the table is nailed to the stakes about six to eight inches under water level. The ducks can then swim in and feed. Yet

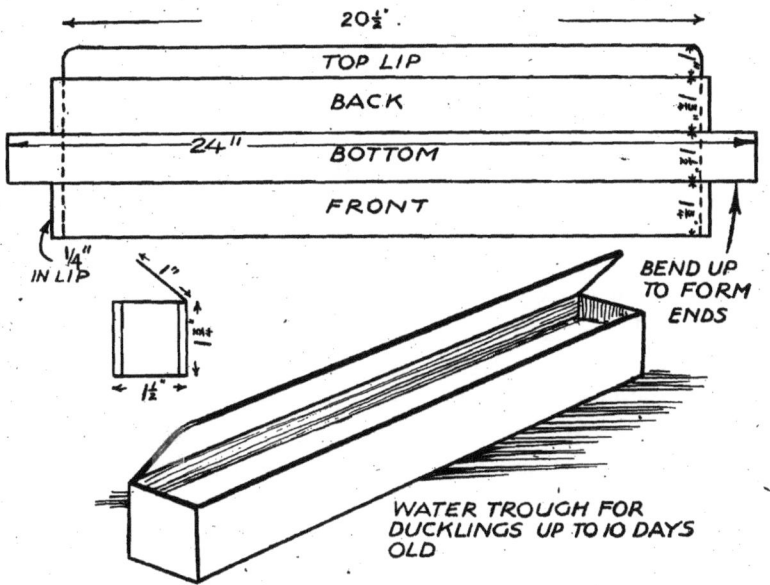

Water troughs can be made on the farm from galvanized tin sheeting. The sketches show general lay-out and the necessary dimensions

another exception is feeding grain direct on to clean grass in small heaps or in a thin line, changing the feeding place each day.

For all other feeding, galvanized iron troughs are to be preferred. They last a long time, are easy to clean and are light to handle, which cannot be said of the cast-iron type of trough. Galvanized iron keeps clean and bright and does not rust for many years if of good quality and galvanized after making. There are on the market some very handy sized pig troughs, and these are very suitable for feeding mash to adult ducks, also the grain feed under water.

TROUGHS FOR YOUNG DUCKLINGS.—The best food troughs I have yet found are easily made from flat, thin galvanized sheeting, and they do not require soldering.

They can be made in different sizes to suit all ages of ducklings from three days to five to six weeks, after which larger troughs can be used. They are best made from two feet wide, six feet long, thin grade galvanized iron sheets—a sheet will make several troughs. The great thing is to have plenty of length of trough room for the youngsters.

Those who want to save money can make excellent wooden troughs of all suitable sizes from odd pieces of board or box wood.

FOR ADULT DUCKS.—An excellent trough is made from galvanized ridging, usually sold in six-foot lengths. Cut in half this gives two three-foot long troughs suitable for from 10 to 12 ducks. The whole six-foot length will serve for about 25 birds; or the length cut up into three will make three troughs, each suitable for a small pen of a drake and four or five ducks.

To make a trough we require some six- or seven-inch wide sawn boards, about one and a half inches thick. Saw off two lengths about 14 inches long for the two ends. Next we require four pieces of two-inch by one-inch batten about six inches long; nail these on to the ends in "V" shape, having the top of the "V" to come to the top of the ends and the "V" to the width you desire. Having your two ends ready, place the ridging into position and nail firmly into place with galvanized slate nails. If the joints are done with thick paint before nailing the trough will hold water. On the other hand, if it is never to be used for water, punch a number of holes in the bottom; it will then not hold rain water if kept in the open.

When rearing large flocks of commercial ducklings for killing purposes at range and in the open, many rearers use empty sacks, of the one hundredweight type, in which mash is delivered. Laid on the ground the wet mash is placed

NATURAL SURROUNDINGS

A flock of Young Aylesbury and Pekin ducks enjoying the full benefits of a running stream. Their journeys are, however, restricted by a series of netted compounds

on them in heaps, and the sacks last quite a considerable time in decent weather.

Others use platforms of wood about a yard square, on which are placed heaps of meal twice daily, early and late. The ducks eat it as and when they like, but are made to clear up each day's supply to avoid sour food.

Both these methods are, of course, for partly feathered ducklings out of doors day and night and with a plentiful supply of water. Both are successful and save a lot of labour.

While I do not subscribe to the view that anything will do if it saves capital outlay, I do feel that a great deal of money is often wasted on unnecessary appliances. We shall always find people who are ready to spend far quicker than they are ready to fill in a spare hour or two in their own little workshops making a few accessories.

I do most of my jobs of this type during the long winter evenings and I have to confess that I get a great kick out of making good an appliance which at first sight appeared to be past repair.

Another money-saver in the long run is to buy only the best. Cheap houses soon become shoddy and my experience taught me many years ago that houses of this type can seldom be repaired.

Chapter VI

THE WATER SUPPLY

SWIMMING water is not absolutely essential for waterfowl, yet it pays to provide swimming water when possible. By this I mean water containers of sufficient size and of a suitable shape so that the ducks may make their toilet. From a health point of view it is worth going to a fair amount of expense to provide this suitable water supply.

Failing swimming water, the drinking water must be given in containers of sufficient depth to allow the birds to immerse their bills and heads.

Ducks without swimming water can easily be provided with serviceable water holders by getting five- and ten-gallon oil drums and cutting them down to about six or seven inches in height. These are easily obtained, often free for the taking, and they certainly make useful water tins about 12 to 14 inches in diameter. Better still, they are easy to replace and one can have quite a few of them on the place. They are best kept in a row, about a yard apart, and moved each day so that the ground around them does not get trampled on by the birds. Do this and the grass remains good.

CEMENT POOLS.—If you have no natural pond or stream and mean to go in for ducks, by far the best plan is to build the desired number of cement pools. They save a lot of time and bother and give enjoyment and health to the birds. They are especially useful if one can arrange for a piped supply of water. Any amateur can make quite nice-looking pools. Made carefully with a proper mixture of sand and cement, they will last a life-time. A useful size is five to six feet in diameter, saucer-shaped and about 12 to 15 inches deep in the centre. It is a mistake from a cleaning point of view to make them too large.

Let us imagine we wish to build a pool for 25 ducks, which are running in an orchard near the house. First, pick on a high spot, the highest in the orchard, yet taking into consideration the fact that we have to get water to the

DUCKS

pool, either by hand or through a pipe. Always remember that the birds can easily walk to the pool, whereas you may have to cart the water to it.

If you pick a high spot and build your pool well above ground level and then make up the ground around the pool, it gives you every chance to arrange a drain from the pool and thus save a lot of labour and mess in frequently emptying it.

It is quite easy to fix a sink or bath plug in the cement pool. Personally I use the top part of a bath plug, and in

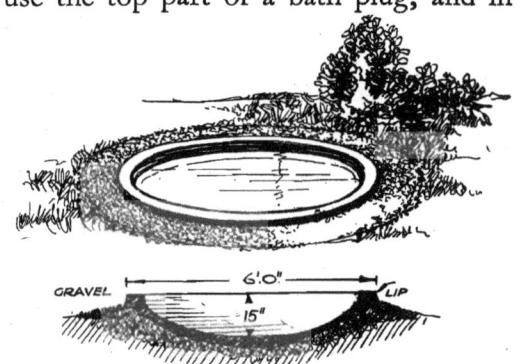

In the absence of a natural pond or stream an artificial pool will give enjoyment and health to the birds. The construction of a pool is described on page 83.

place of the bottom attachment I screw on an iron elbow and to this connect some old iron piping—this is, of course, from the centre and under the pool—then when one wishes to empty the pool one merely stirs up the water and pulls out the plug. It is wonderful what labour can be saved in watering if one can run three-quarter-inch piping, buried nine inches under ground, to the different pools. If by good fortune you have a supply of water, either town or from a gravity tank, you can quickly run water by this piping to many different pools. Do not cement your supply pipe into the pool, or you will have to chip your cement to get the pipe out if anything goes wrong. Far better to have a tap at the end of the pipe and overhanging the lip of the pool.

Having picked a suitable situation, mark out a circle of the desired shape and pare off the turf. When marking out, mark it at least one foot larger than the pool will be when completed. Now dig out the soil to a suitable depth and shape, allowing for, say, six inches thickness of concrete and the finish of cement and fine sand. Next ram the soil firm and if you have some broken stones or shingle place a layer of this and ram it firmly. Old pieces of wire netting are also an asset when placing cement in position.

Personally I do not use concrete when making a pool. I work with a mixture of fine shingle and cement, one part cement and three parts shingle and sand. I am fortunate enough to have an old iron hoop from a wheel, about five feet in diameter and six inches wide. This is placed in position, and if necessary propped up on bricks or stones so that it is level and has its edge about five inches above the ground. When all is ready I put in the cement and work it into shape and position, well tamping it—saucer-shaped and coming gradually up to the edge of the hoop. Having got this to my liking I leave the whole for a few hours, then, when I consider the cement to be set, I carefully remove the hoop. More cement is then used to make an edge, about 9 to 12 inches wide, right round the saucer. So as not to use too much cement mixture I keep packing in clean washed pebbles and stones. The edge of this lip is kept fairly upright.

The whole is now well polished up, especially in the saucer; rough markings are made on the lip and the whole is covered over for the night and as soon as the cement will stand it, water is frequently poured on with a watering can.

If it is wished to fix a plug in the centre, simply place a pot or glass jar in the centre of the pool when making it; later, this may be taken out and, when the cement is hard and dry, tunnel under the pool and place the plug and pipe in position and cement the whole in with a mixture of one part of cement and one part fine sand.

Around the lip of the pool clear away more soil and fill in with stones, clinkers or gravel, ram well down and finish off with fine shingle. By doing this it will be found that the ducks do not trample much dirt into the water, nor do the surroundings become messy.

These pools also prove ideal for ducklings when they

get to three weeks and older. When you have some ducklings using these pools, go to any old pond and collect a bucketful of duckweed. Place this in the pools and watch your ducklings, and you will agree that pools are well worth while.

PONDS, STREAMS, ETC.—Much can be done to improve existing ponds or streams so that ducks can have good water throughout the year and especially during hot, dry times. The time to do any such work is during the summer, when the water is low.

Clean off the banks, dig out the mud, make good the banks where necessary, and make suitable entrances and exits. This is often best done with concrete.

Even very small streams may be made so that the water from them will give hundreds of ducks both swimming and drinking water. Dig out pools at different points, then place a small dam in the bed of the stream. It is usually possible to divert some of the water, run it down a cement channel, and then return it back to the stream lower down, after making it serve its purpose for lots of waterfowl. One never misses the water until the well runs dry, and it is well worth while to endeavour to become the owner of an adequate supply of water.

WATERING DUCKLINGS.—Each year thousands of ducklings come to an untimely end, or are so checked that they are not profitable. A duckling, newly hatched by a hen or an incubator, is in no way different to a chicken of the same age, other than in shape and that the duckling will know how to swim by instinct and the chicken will not. The duckling has no more oil in it or on its down or fluff than the chicken and if you allow the duckling to get into water it will get very wet, really saturated, wet and bedraggled. By instinct it will make for and use water, but remember that you have gone away from nature by hatching with hens and incubators. Even a hen-hatched duckling can stand the effects of water better than one hatched in an incubator.

You may have seen tiny ducklings swimming and even chasing flies over a large pond. Correct! But they were not hatched under a hen or in an incubator. What you saw were hatched by a duck, and were most likely 60 hours old and only just on the water.

Generally speaking, ducklings which are naturally hatched under ducks have received a coating of natural oil from their mother's body and feathers and are thus water resisting. They as good as float on the top of the water, not in it, and the water does not get down to the skin. Watch such ducklings come ashore or get on to a large lily leaf—a good shake and all the water is gone! The ducklings hatched in an incubator or under a hen would get wet through and quite " sloppy "—until Nature gets to work and the duckling gradually receives its natural oil.

However, all ducklings must have water in plenty—never allow them to get thirsty and never give icy cold water direct from a deep well. If well water is used, take off the chill or draw off a tankful and allow it to get the chill off.

Never supply water in shallow, saucer-shaped containers—the ducklings love it, but the water is quickly on the birds' bodies and on the ground, making an awful mess. Worse still, the youngsters are quickly without drinking water and very thirsty, and when next their attendant does give them a supply they nearly burst their crops—which brings us to the following simple but vital point in successful rearing: never feed ducklings until they have had a drink. If, by a misfortune, the water-pots are dry and the ducklings hungry, do not feed them but give water, wait a short time and then feed them. By these means you will avoid owning a lot of little undergrown ducklings which are in shape not unlike a small balloon, with a small bill and head attached. Extended crops, all wind and indigestion, no growth, and a very early death.

The remedy is to have the correct size and shape of water-pots and plenty of them; water containers of sufficient depth so that the ducklings can immerse their heads and yet not get half-drowned. It is essential that some of the containers are of good depth, as, apart from drinking, the ducklings must be able to keep their bills, eyes, etc., bathed and clean.

DUCKLINGS WITH HENS.—If you have, say, a hen with 12 to 15 ducklings in a coop, use small round tins cut down to the required height. For the first few days tins about 2½ to 3 inches in diameter and about 1½ to 2 inches deep will serve admirably. It is quite easy to let the tin or jar into the ground if necessary. For the first few days drop a few clean pebbles into them so that the ducklings can get

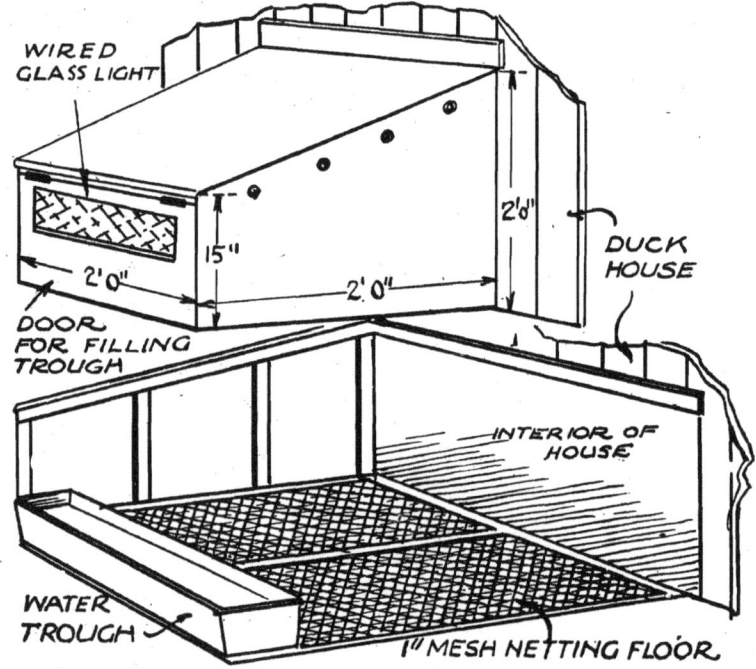

A suggested attachment to a duck house for those who house their stock. It helps to keep the house dry, as the ducks have to walk on the wire to the water and water spilt from the trough falls to the ground

drink and yet not get into them to get wet. As the birds grow, use larger containers and keep moving them further away from the coop front, the further away the better, as it gives the ducklings more chance to get rid of spare water and one can thus keep the coops drier.

Once they are 15 to 20 days old and if the weather is good, there will be no harm in allowing an occasional bath in a good big container, say a five-gallon oil drum cut down to three or four inches in depth; this is both beneficial and healthful.

INCUBATOR-HATCHED DUCKLINGS.—Here, unless care is used, we are rather up against it, as usually these birds are reared in groups of 50 or more. (See "Artificial Rearing.") Water must be supplied in properly shaped tins and troughs, and these are best made from galvanized iron sheeting, for the first week about 18 to 24 inches long, 1¾ inches deep, 1½ inches wide. This gives plenty of drinking space, especially if two troughs are in use with each group.

As the youngsters grow, larger troughs are brought into use, such as 24 inches long, 2½ inches deep and 2 inches wide. Once the ducklings are drinking well it will be found best to arrange that the troughs are behind a grid through which the birds put their heads and necks to drink. By these means it is possible to use troughs of good width and depth so that the ducklings can immerse their heads and yet not waste water and make their surroundings too wet.

Give a group of 50 ducklings all the water they want in open and too large containers, and what happens? They get saturated, cold and chilled, and then crowd round the lamp or in the hot compartment, resulting in a bad chill, colds, pneumonia and deaths. Even the survivors are only useless specimens, unfit to carry on their race. Getting the right water containers certainly simplifies duckling rearing.

Once the ducklings are over the baby stage, and are running on land or wire floors, they quickly "get up" some natural oils and will soon derive more benefit than harm if they have large water-pots into which from time to time they can take a bath. Later, they can go on to natural water, or cement water pools, so long as there is a proper and easy entrance to and from the water.

Arrange an easy entrance to the pond or stream, gradually move the water-pots nearer and nearer to the water entrance, and very soon the youngsters will be carting their own liquid refreshment. This is much the best way to introduce ducklings to a pond or stream.

Chapter VII

SELECTING BREEDING STOCK

IN selecting and mating breeding stock both the hands and the eyes must be used. It is also of great help if one knows how the birds have been bred, and what they have done in the past—what sort of performance they have put up as layers, size of eggs, colour, texture, shape, winter production, and many other useful points which one may have kept in mind or in a note-book.

Generally speaking we may conclude we are working on birds which have produced for one year, and which are to be used as breeders to give us our future producers, or, if table breeds, to produce both table ducklings and future breeders.

Having got the ducks handy and ready to pick over, what exactly are we going to look and handle for? First, breed type and characteristics, meaning shape and colour as laid down in the breed Standard. Next we must handle for many things. With two or three fingers of one hand between the legs, the other hand on top, the bird should feel " balanced."

It should have tight, silky, glossy feathering, really waxy and capable of keeping out a cold east wind: not long, harsh, woolly feathering which lacks oil and will let in cold; you can feel the difference easily.

What you want to feel is a bird which is well made and solid, not all feathers and no inside. Choose birds with good strong bills, well set into a refined skull, bright eyes, well-placed, medium-to-thin neck, a long, wide back, with plenty of heart and lung space. Throw out any duck which has the least suspicion of hump back; the back must be level—flat right down to the " parson's nose."

Select birds of good size for their breed, not small ones. On the other hand, throw out all of the coarse, heavy-browed type, heavy in bone and generally with the class of harsh, long feathering already mentioned.

I like to see each bird "on the move," no matter what the breed; action counts in all stock. The duck which carries herself properly and "goes well" is the sort likely to throw good progeny. The unsymmetrical, heavy-boned, thick-necked and heavy-headed, sleepy sort will assuredly let you down.

Working with ducks each day, you will quickly know the type which produces and does her job; also, you will have a good idea of the breed Standard, and with this combined knowledge you should be able to choose your best birds as future breeders. It always pays to use a few of the very best as breeders in preference to numbers of birds of which you are not quite certain, as most likely you are simply letting yourself in for trouble and disappointment in the future.

In selecting and mating stock one must use some judgment as to how many ducks to mate with a drake. This, of course, depends on the breed, age and surroundings. For instance, if a big breeding flock of ducks have absolute free range—say a breeding flock of Aylesburys—it is possible to use quite a high percentage of drakes early in the season, say a drake to four or five ducks. The birds split up on the range and thus the drakes do not interfere with each other, as they would if in a confined pen.

Let us run through different matings. In the first place, remember that one can mate safely in pairs, if wished, or even a drake to two, three or four drakes in any breed. In heavy breeds such as Aylesburys and Pekins a drake will mate with two, three or four ducks, or, if they are very active utility-bred birds, with five ducks. In all cases when using more than one male in a pen or flock it is wise to have brother or friendly drakes which have run together before the breeding season arrives. Two drakes mate with 10 or 11 ducks; three drakes, 15 to 18 ducks; four drakes, 25 ducks. When using an old drake who sires very fine birds, just give him, say, two of your very best and most suitable ducks, or even one.

In light breeds, such as Runners and Campbells, a drake will take any number up to six or seven ducks; two drakes, 15 ducks; four drakes, 30 ducks. In the medium-sized breeds such as Orpingtons, Cayugas, etc., a drake and four or five ducks is about the best mating.

Generally speaking, it will be found that the average drake is fertile from March and onwards; some drakes seem incapable of fertilising an egg although they mate all right with the ducks.

One is often asked whether drakes should be taken away from the ducks when the hatching season is over—is it harmful to leave them with the ducks? Personally I always leave the drakes with the ducks throughout the year—often for a number of years. A drake will only mate with a duck when she is near laying or in lay, so that with breeding stock, which is generally not encouraged to lay in the winter or too early in the New Year, the drakes are not active.

I would recommend mating up the birds and leaving them mated until you require fresh drakes. In a wild state most ducks mate as a pair and even now, after all these years of domestication, a duck and a drake of any breed will mate and live peacefully together throughout the year.

In the smaller breeds such as Black East Indians or White Decoys I prefer a mating of a drake and, say, three ducks. However, a Black East Indian drake will take up to five ducks, whereas the Decoys are best in pairs, or a drake and two ducks.

Chapter VIII

TRAP-NESTING

FORTUNATELY, ducks lend themselves very well to trap-nesting. They very quickly settle down and in a short time will trap themselves each evening. Unlike pullets, which lay during the day, ducks lay during the night, and it is not really costly to trap them, as the trap-nests can be arranged so as to be in the house in which the ducks sleep. It really means that each duck is housed separately in a small pen. To make it more convenient a trap is fixed in the front of each compartment.

It does really pay to trap ducks, as by doing so any owner will quickly find out many important things. Small eggs, misshapen eggs, ducks which lay soft or porous shelled eggs, poor layers, or the duck which looks as if it is in full production yet never lays—all these are discovered by the trap-nest.

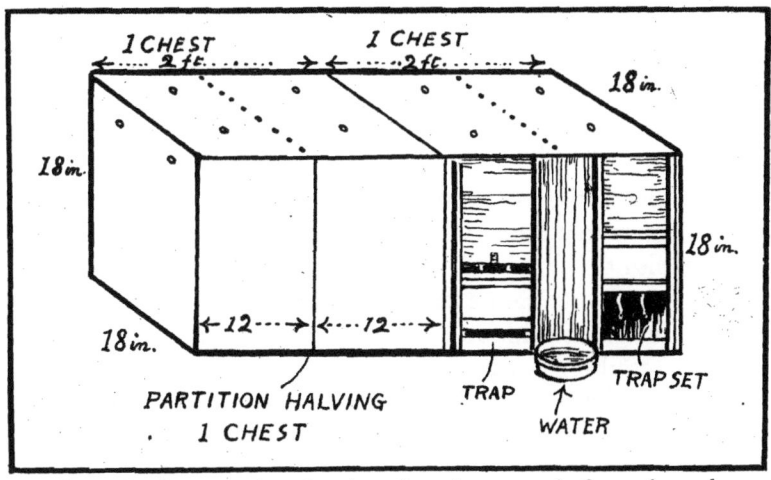

The author's suggestion for cheap trap boxes made from three-ply chests; they are suitable for use under a shed or in a house

It is just as cheap to house breeding stock in trap-nests as in a duck house. A handy, neat and light trap house can be made easily and cheaply for a drake and his ducks, each in its own small " bedroom," no pushing and crowding yet in view and hearing of each other.

I would put the limit of a successful laying flock sleeping in one house at night at 50; 25 will do better, 12 better still. There are always highly strung, nervous ducks in a flock, birds that are always on the move and never settle down,

TRAP-NEST FRONTS
Here is a pair of the orthodox type of hinged front, showing one on the left sprung ready and the other closed

and this habit passes on to less nervous birds and in some cases to the whole flock, which is always on the move throughout the night.

With a suitable range and, say, the use of an old roomy building which could be fixed up as a trap house, it would be quite feasible to run a flock of hundreds, for the simple reason that they would each have a separate little house at night. It has been tried and works well with 250 or more in a flock. Any wild, scared, nervy individual gets into one compartment, and thus will not affect others or the whole flock.

DUCKS

Any small owner, running ducks as a hobby, will get an added interest by trapping each season's young birds, taking the cream of the year's breeding, getting them tame and used to the traps in the late summer and keeping proper records. For myself, if working with only a small number of birds I would " house " all my birds in traps.

Home-made trap-nests are easily built, or you can buy the whole equipment ready for the ducks. If you only run a few ducks, a number of three-ply chests can quickly be converted into grand duck traps !

As stated in an earlier page, I fill in many winter evenings

SUN PARLOUR DUCKLINGS

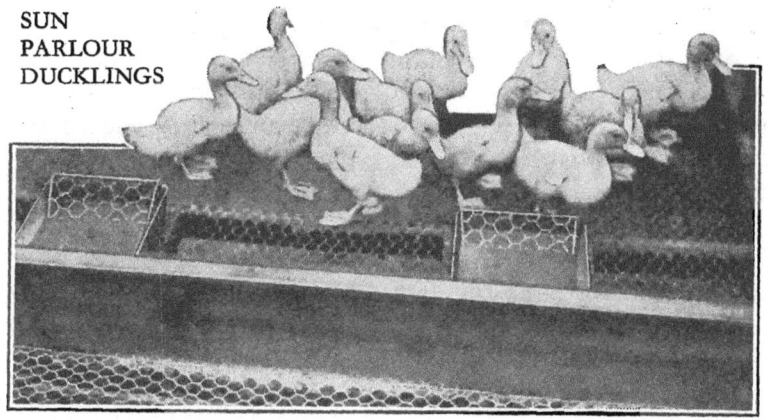

This photograph shows a novel form of drinker employed for ducks in the sun parlour of an intensive house. The vessels are deep enough to allow for complete immersion of heads while the wire-netting attachment prevents drowning

repairing and replenishing appliances. Renovating trap-nest fronts is only one of the tasks. One of the main points concerning them is to see that the slide works freely in its wooden grooves ; this is often a question of correct weighting.

Chapter IX

EGG COLLECTION AND INCUBATION

COLLECTING AND WASHING.—All eggs should be collected each morning, after liberating the ducks from their house or compound. They should not be collected in a dirty bucket and left there—eggs quickly take on a taint; if you must collect them in the food bucket, line it with hay or straw. It is a very good plan to wash eggs every day, the sooner after the collection the better; the dirt is then more easily removed.

Soak the eggs for five minutes or more in water which has just had the chill taken off. A good sized, hard-bristled nail brush should be used. Remove every particle of dirt and marks, and place the eggs on one side on a wire tray to drain; but be certain to dry them really well before they dry on their own.

The secret of success with ducks' eggs is to get them really clean; it is absolutely useless to market dirty eggs or only half-cleaned ones; if you do so you are spoiling both your own trade and that of other people.

If eggs are for incubation, on no account must water which is too hot be used; if reasonably clean, just wipe them over with a damp cloth and finish with a dry one. It pays to make them clean for incubation purposes, as a clean egg is easier to test later on for fertility.

STORING.—For eating purposes all eggs should be sold before they are seven days old. In any case, eggs should definitely be cleared out once in seven days. Eating eggs must be stored in an even temperature, preferably in a dark place; on no account must they be near any strong smelling articles.

For hatching in incubators, put the limit of storage at seven days. It is best not to keep them longer than three or four days before they are in the machines. If they are going under hens they can be kept for any time up to a fortnight. In either case they must be carefully stored on trays; it is immaterial whether they are on end or on their sides (I have never yet seen a bird leave them on end in a nest!) The trays should be bedded with clean-smelling chaff, cut hay or cut straw. A cool, even temperature is essential, and keep the eggs covered with clean sacking. Turn them each day.

TESTING.—Every egg sold should be tested. By this I mean the egg should be placed before a testing light so that no bad specimen goes on to the market. On no account let an egg which is cracked, has a blood spot or any other spot, a broken or watery yolk, etc., get on the market. It does untold harm to duck egg sales.

Test every egg before it goes into the incubator or under a hen. It is wasting space, eggs and time to try to incubate a faulty egg. Be very careful as to shell texture, colour, etc.; remember that if such eggs do hatch, it is practically certain that the resulting ducklings will carry the fault, and if used later on as breeders, will again pass on the defect. On the size, shape, colour and texture of the eggs used for hatching purposes depends the results in good or second-class eggs in the future. Have only the perfect egg, using both hand and eye and remembering what class of drakes you are using as sires and from what class of egg they were hatched.

IN THE INCUBATOR.—Taking the incubator tray out of the machine, place it on a table with the front towards you. On each egg place a "X" one side and an "O" on the other; if special eggs, instead of the "O" write "WR" for White Runner, "KC *1*" for Khaki Campbell, Pen 2, and so on. If from trapped ducks, later to be hatched in pedigree trays or bags, put the number, etc., in place of the "O".

It often helps to use coloured pencils. For instance, there may be six eggs from KC 980—put a red ring round the egg; five eggs from KC 986—put a blue ring round, and so on. This saves a lot of sorting out when they are ready to go into hatching bags or pedigree trays.

Commence to tray the eggs by putting two or three in the centre of the tray with the thin ends towards you. Now place some on each side until you have a complete row across the tray. Then begin in the middle of the row, and place more eggs until you have another row. Continue to place the rows, all with the thin end towards you and finish off half of the tray in this manner. If there is only room for about half the width of a row of eggs at the front edge of the tray, gently work the eggs back so that you can place another row on. Next, turn the tray completely round and put on more eggs, but this time remember to place them with the large end towards you. You should finish with the eggs evenly and well packed. This will help you a great deal throughout incubation, testing, turning, etc.

I give the following tip for what it is worth and the reader must please himself whether he makes use of it. Place 10 to 15 eggs down one side of the tray on top of the other eggs, and each day change them with others which are on the tray (as the top eggs are getting one degree more heat). These eggs come in very useful when we test at the fifth day, to replace clears, etc.

The method I have given is much better than placing eggs anyhow on the tray. They are in perfect rows and are easy to turn. If you do put eggs on top of others, for the first five days be careful to see that they do not hit the capsule stirrup; also that they clear the thermometer.

SELLING.—It is folly to expect somebody else to sell your duck eggs for you! Do not wait for customers to run after you demanding eggs. In the first case you will have to pay the other party for his work and in the second you may have only a few coming to you for eggs. If you produce an article it is up to you to produce the right quality—grade it well and place it on the market in perfect condition. Don't run round saying you cannot sell your eggs—that there is no demand for them, etc. Find a demand or market for them and remember that those who buy your eggs wholesale have to sell them at a remunerative price to the public, or they would not take your eggs.

When you have made a market, keep it by supplying the perfect article, clean and up to a proper sample. If you live where there is not a good sale, send them to a district where there is. As a matter of fact, one can sell duck eggs in any district if one supplies the goods. Give some away to likely people—let them try a real, properly produced duck egg, and they will be your regular customers. Duck eggs will make the same price, often more, per dozen, as the best first-grade hen eggs. If you have both green and white eggs, mix them 50–50; they show each other off. In certain districts a sample of mixed coloured eggs sell the best; in others all white eggs are demanded.

INCUBATION

Under this heading I propose to deal with all the useful methods practised to hatch out strong ducklings. In many cases the owner of a flock or pen of ducks only wishes to hatch a certain number of replacements each season, and broody hens will serve such breeders. In other cases a larger number of ducklings may be required, to give replacements and to increase the head of stock. Then we have those who may wish to hatch out quantities for the farm, for sale as ducklings, or stock. In these two cases it would be next to essential to use incubators.

HATCHING BY BROODY HENS. For the best results broody hens cannot be beaten, for they hatch out a big percentage of ducklings from fertile eggs. They can give something which we as yet do not know how to provide, and this in the first ten days, when the germ in any egg requires care.

If you decide to use broody hens you have a mother ready for the ducklings when they are hatched. First procure some good reliable broody hens of medium or heavy breed. Do not be afraid to use pullets; in 98 per cent. of cases they sit well and have the big advantage that they are not so heavy as hens. Fix up proper accommodation as a broody house, a good rat-proof place with a sound floor, reasonably lighted and airy. Have each hen under control, in a nest which has a door to it.

Two- or three-compartment orange boxes will prove satisfactory. Just leave on one piece of the top, turn the box on its side and the piece of wood will be there to retain the nesting material. Then fix doors so they come to within an inch of the floor and hinge them so that they will turn back

IMPROVISED NESTS
Natural methods, using hens as broodies, are still widely practised even on the larger duck farms

on to the top of the box. To each nest allow two bricks to keep the door fast when the hen is on the nest. When the hen is liberated for feeding one brick is placed on the floor in front of the nest to form a step, the other being put on the door, which is turned back on the top of the box.

For nesting material cut sods or turves about $\frac{1}{2}$-inch larger than the bottom of each compartment. Place them in grass upwards, and beat them into shallow saucer shape. If too flat the eggs get all over the place, but if it is made too concave the eggs are always pressing on or getting on top of each other.

Treat in and around the nests with a reliable insect powder

and creosote the whole box inside and out about a week before using. Then, with a little sprinkling of cut hay, cut straw or wheat chaff, we have a good nest.

Provide each broody house with suitable food and water pots, and grit in a box. It also pays to provide in a box about 2 feet long, 18 inches wide and 1 foot deep a mixture of fine dry earth and ashes in which is some flowers of sulphur, insect powder, etc. The hens love it and use it regularly. The hen which is nearly eaten alive cannot make a success of her job, and in her discomfort she will smash many eggs.

Every other day cut and give the hens a turf of green grass. Sweep up the broody house each day and sprinkle fresh sand or sawdust on the floor; this makes things more pleasant for the attendant and for the hens, and keeps the eggs cleaner and easier for you to test for fertility, etc., later on.

Make an effort to sit hens in groups; it saves time and trouble during incubation and, later on, in rearing. Test all hens for two days before trusting them with good eggs. Once you give the hens the eggs, place on each nest a card with such details on it as, " date sat," number of eggs, breed, etc.

Each day, as near as possible at the same hour, allow the hens off for from five to fifteen minutes. For the first three or four days allow them just a short time and then increase up to fifteen or even twenty minutes towards the twenty-eighth day during hot weather. Personally I find it best to open the nest doors and carefully remove the hens which do not come off on their own. It is best to use one's own judgment as to whether the nests are left open or closed during feeding times.

The advantage of using grass turves is that you have a nest which is reasonably natural and in fair contact with the floor; also, it will absorb some moisture, drawn up through the turf and to the eggs by the heat of the hen's body. In dry times moisture can be applied by pouring water on the floor at the back of the boxes.

Once you see an egg chipped, or hear a duckling, the hen should remain unfed or watered until she has finished hatching. If wished, any empty shells can be carefully removed. Some people like to remove the early ducklings, keeping them in a warm, flannel-lined basket near the fire. Those who have an incubator running can of course place any ducklings in it and thus make certain they do not get trampled to death.

Feed the broody hens on maize and just a little wheat, clean water, grit, grass turf or greenstuff. Try always to keep a spare hen sitting on clear or dummy eggs, ready to replace any hen which becomes unsteady or runs the nest.

Just one thing more. At certain seasons of the year broody hens are difficult to obtain and a good tip is to make arrangements with a large commercial egg farmer to have some either on loan or at so much per head, to be returned when they come into lay.

My own plan is to run up to twenty hens in each broody house, trying to sit friendly hens in groups, and liberating three to six and often nine hens at one time.

HENS COMBINED WITH INCUBATOR. This scheme has much to be said for it in the results given. The idea is aways to keep a certain number of hens sitting on eggs, the number depending on incubator capacity.

The hens sit on each lot of eggs for ten days; they are tested at about the ninth day and the good eggs are placed in the incubator. More fresh eggs are given to the hens for another ten days, tested and again put into the incubator. Then one uses one's judgment, either giving a third lot of eggs or allowing the hens to hatch out their first lot, which are taken from the machine. It really means that each hen will incubate three lots of eggs each for a period of ten days.

It is a positive fact that the duck or hen can beat the incubator, and it is during the first ten days that she does it! By this time the fertile eggs have partly formed ducklings in them and with good management they will carry on all right in the incubator. By this method we have mothers ready for the ducklings when they are ready to leave the incubator. Another thing is that one may, if anything unforeseen crops up, allow the hens to have the eggs for fourteen days or to hatch them out if necessary.

DUCKS

HATCHING BY INCUBATOR. No matter what the size of incubator, one can reckon on its capacity as being about 85 to 90 per cent. when using it for duck eggs; even then it of course depends on the size of the eggs. A 150-egg machine will take about 130 large Aylesbury eggs, or 140 Khaki Campbell or Indian Runner eggs.

Much has been written on the incubation of duck eggs, lots of it largely concerned with different temperatures, humidity, atmosphere, etc. I do not, however, intend to go into a lot of detail. I propose to make an effort to explain clearly how best to get good, strong ducklings.

The secret of success is: Fresh fertile eggs with good sound shells laid by matured stock, properly fed and cared for. Without such we cannot make a success of incubation, natural or artificial.

Have the machine running at an even temperature of 103 to 103½ degrees F. Tray the eggs (as advised in the previous chapter). Place the tray of eggs in the incubator, but leave the door open. It is best to place the eggs in the machines about 7 p.m., leaving the door open until the last thing at night, then closing the door. If you place a tray of cool eggs in an incubator running at 103 degrees F. or over you will certainly get a few burst yolks—and it costs nothing to leave the incubator door open for about three hours until they warm up a little.

Follow the incubator maker's instructions as near as possible. See that the felts are in position and, from the start, keep the water trays filled. Having shut the door at, say, 10 p.m. there is no need to do anything until the next evening, when the tray should be taken out and eggs quickly turned, meaning that each egg is turned over so that the underside becomes uppermost. Do not cool for the first seven days but turn each morning and evening. At the morning operation place the tray back in the same position, but each evening reverse the tray, so that if there is any variation of temperature in any part of the machine, all the eggs get a share of it.

On the seventh day test the eggs for fertility (see " Testing

Eggs "). Try to keep the machine running at an even 103-103½, although 104 degrees will not matter, especially towards latter part of the hatching period.

Now we come to the question of moisture. For the first seven days apply none direct to the eggs, only keep the moisture tray or trays full. During the first seven to ten days avoid chilling the eggs, simply turn each morning and evening.

When testing, do not try to do it all one evening, take two or three evenings. On the eighth morning, about midday, sprinkle or spray the eggs with warm water; do it quickly and close the door, and repeat the operation on the tenth, twelfth and fourteenth days. Then, depending largely on the season of the year and what your incubator room is like, give a sprinkling or spraying every other day until the first " cheep " of a duckling is heard.

When the first " chipped " egg is seen, or the first duckling is heard, get a basin of hot water, heated so that you can just bear your hand in it, give the eggs a really good wetting, throw the remainder of the water on to the nursery tray canvas, close the door and leave everything alone until hatch is completed. If for any reason you have to open the door, do it, quickly. Try to keep up the humidity in the machine.

It is best to keep the ducklings on the egg tray until a good number are hatched and really dry. To do this, roll one of the felts and place it in position between the window of the door and the edge of the tray. From time to time, if wished, one may quickly let down the dry ducklings, but do not let them down until you have a good number, so that they can help keep each other warm on the nursery tray.

Fill the oil container every week, and have a fixed day for the job. Commence each season with a new, dry wick and clean the lamp and wick each day after you have turned the eggs each morning; do not trim the lamp and then turn the eggs with oily hands. Each season have the thermometer properly tested, by a chemist if you do not know how to test it with a clinical thermometer. Fix your thermometer in the machine so that the bulb is just a fraction above the eggs. In very dry times stand a bucket of water under the machine and sprinkle water on the floor.

With hot-air machines there are two felts which go on the nursery tray. Commence with both in position, and on the seventh day remove one. The instructions given by most makers are to remove the remaining felt for the final week, but it is best to work this out to mean removal on the nineteenth evening. When starting the eggs in the incubator, place a card on top of the machine with the following particulars: When set; remove first felt; remove second felt; date due, also any other details of use and interest.

Never remove any ducklings from the machine to give to hens or to place in rearing equipment until they are thoroughly dry, active and absolutely strong on their feet.

TESTING FOR FERTILITY

Duck eggs are fortunately quite easy to test by the light of a testing lamp or before any light (in a darkened room), by holding the egg between the first finger and thumb of one hand to form a shade around it and turning it with the finger and thumb of the other hand. By so doing one is able quickly to note what is in the shell or any faults there may be in its texture. In all cases, due allowance must be made for the colour of the shell.

By looking for faults in the shell formation one saves many eggs which would only go wrong during incubation whether in an incubator or under hens. At the same time, you are helping to improve shell texture of the eggs laid by your future stock.

For all general purposes when once you have tested the eggs and placed the good ones under hens or in the incubator, it is best to leave the next test until the seventh day. The germ in any egg during the first week does not require much chilling to cause it to die, so that I would advise all but the expert to leave testing until the seventh or eighth evening. Then can you test, but avoid chilling the eggs.

An ordinary small electric torch is the best light to use when testing at night, or in a dark room.

Holding the egg between the first finger and thumb, place the torch light underneath, and if necessary, move the egg about a little. You will be able to see immediately what progress has taken place in the seven days and act accordingly.

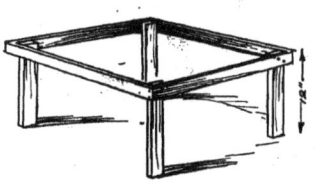

An easily made stand on which egg trays can be placed when testing is carried out

When testing eggs in incubators, I place the tray on a table and run the torch light along under the rows of eggs, if necessary moving them with the forefinger of my other hand. One can detect the "clears," broken yolks, dead germs, etc., at a glance. Pick the eggs off the tray and place them in a suitable receptacle; they can be run through again later should you be at all uncertain about any of them. The great thing is to act quickly and get the good eggs back into the incubator.

When testing in a hot-air incubator with a single tray (say 100- or 150-egg size), if you have trayed the eggs properly in rows, a very good and quick way is simply to draw the tray half-way out, running the torch under the rows and taking out any "duds" with the other hand, placing them on the top of the machine, to be run through later in order to make certain. Then reverse the tray and test the other half. Later, with fewer eggs in the machine you can test again from time to time.

If you are slow at testing, do only half at the first test, and the remainder the following evening. When testing, fill up spaces left by the removal of clear and otherwise faulty eggs by putting in their places those already tested and found to be good. This prevents eggs rolling about and also keeps them in rows, which is most helpful throughout the incubation period. If you are doing only half the tray, just leave an empty place or two in the last row tested.

If you prefer the safer way of testing on a table, it will be well worth while making a wooden appliance, so that the tray will rest safely on it and allow you space to work with the torch underneath.

The accompanying sketches are intended to show what one may expect to find when testing duck eggs.

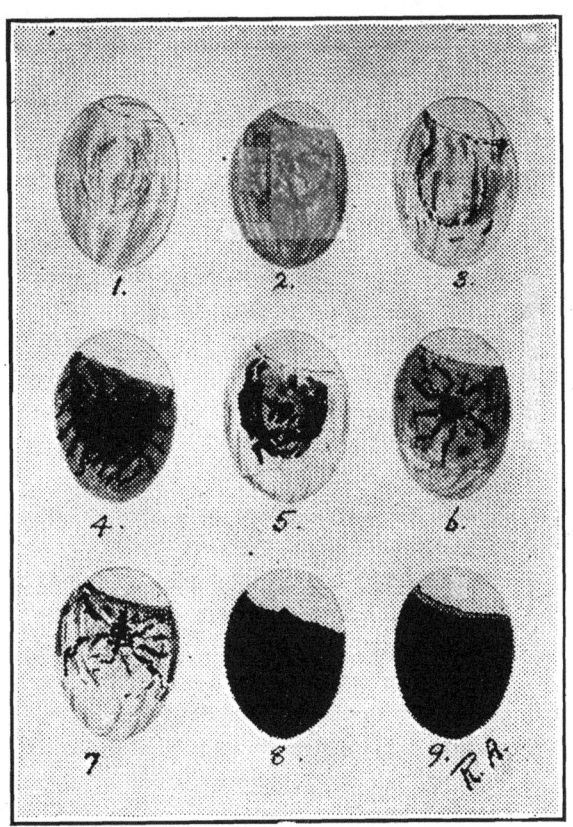

THE EGG DURING INCUBATION

No. 1 shows a clear egg at the sixth or seventh day of incubation. The air-space at the top shows clearly and a little larger than that in a fresh egg, owing to evaporation. The yolk is a little bit heavy and cloudy and can just be seen. Otherwise it is much as a fresh egg would appear—with no germ and no life.

No. 2 depicts a blue or green egg at the fourth or fifth day. The air-space is showing a little larger than in a fresh egg, the round shape of the yolk being clearly seen with the spidery form of a live, growing germ on it. The whole is clearly defined and alive.

No. 3 is an attempt to show a pale green egg with a burst yolk and the red circular line which goes with such eggs, plus a certain amount of " fogginess."

No. 4. Here we see a good egg at, say, about the fourteenth to sixteenth day of incubation. The dark body of the partly formed duckling is seen, also the big blood-vessels; the air-space is about right, showing that the correct amount of moisture has been given.

No. 5 is that of an egg after about twelve to fourteen days, with what was a live germ, but is now dead.

No. 6 depicts a dark blue shelled egg on the seventh evening. Here we have a strong, big, active germ with good " legs," one which is very lively and jumps about when you move the egg before the lamp. The air-space, if anything, is showing rather too big.

No. 7 shows a strong live germ in a pale coloured shell: very strong and lively with good definite " legs."

No. 8. Here we have a good egg as seen on about the twenty-fourth day. It has about the correct air-space, and the edge is very dark and very definite.

No. 9. This is the class of egg which gives trouble to the novice. On testing, it would appear to be a good egg, well filled with duckling. The sketch shows such an egg on the eighteenth to twentieth day; the air-space is too small for the time the egg has been incubated, and the line around the air-space shows a yellow tinge of colour, not a definite dark line as seen in No. 8. The duckling is dead. Test very carefully and, when assured that this is the case, such specimens should be removed. As I have said, with duck eggs it is best to test on the seventh day (evening) for clears, burst yolks, and for germs which have started life but died. Then test again on the fourteenth evening for germs which have

DUCKS

died since the first test. Then again on the twenty-first day for further bad eggs.

When working an incubator, make use of your eyes and your nose, and you will be able to remove any offensive eggs. Make due allowance for differences in texture and colour of shells. Test quickly and regularly; it is most harmful for good eggs to be in contact with, and " breathing " the bad gases thrown off by bad ones.

Bought eggs are usually offered on the following terms: (*a*) A certain price per dozen or per hundred with " clears " replaced within fourteen days, (*b*) a certain price per dozen with three extra eggs making fifteen to the sitting, or, in the case of a hundred, an extra twenty and no replacements.

Thus, if you pay for twelve eggs, with clears replaced, you are allowed to return any clear eggs and to receive fresh ones in their place. A clear egg means one without any signs of a germ in it. On the other hand, if you buy fifteen eggs to the dozen you take your chance and receive no replacements.

I believe, however, the law holds that you are entitled to something for your money, meaning that if you take fifteen eggs without replacements, you are entitled to some good eggs. In any case, most sellers would certainly make such failures good to the purchaser. They are, after all, not of the here to-day and gone to-morrow type. One regular customer is worth half a dozen casual buyers and most of the breeders I know are always ready to boast of the numbers of regular clients they possess. A dissatisfied buyer seldom comes back a second time.

CHAPTER X

REARING METHODS

NATURAL REARING.—It is seldom that I use ducks to rear ducklings, but they often make a great success in hatching ornamental waterfowl eggs, such as Tufted Duck, Pochard, Gadwell, etc., ducklings which seem to thrive better if they receive a coating of natural oils from the duck and can thus very quickly use a water pool, from which they seem to get what is difficult to supply by hand. It means that the youngsters can work on a stream and return to Mother Duck in the coop when they wish, and this they do frequently. However, for all general purposes, I prefer a good broody hen or a bantam hen for the smaller breeds of ducklings—they appear to brood the ducklings more and better than ducks.

Let us imagine some hens have hatched some ducklings. The youngsters are in the broody house, nicely fluffed out, and have been toe marked. There are three hens with 12 ducklings each. They are of a number of breeds and varieties; some of the ducklings are Black Cayugas, others White Runners, some Fawn and White Runners.

They would certainly look nicer if we gave each hen one variety and colour! But don't do it, it will lead to trouble later on, when they are strong and running about; a little black duckling will one day enter the coop belonging to the hen with the white or fawn and white ducklings and the result will be a dead duckling. On every occasion mix up the colours so that each hen has a few of each, then later any hen will take any colour of duckling.

Choose a clean, level, grassy spot, rake it over, and get out the coops; these should have been well creosoted and lime-washed some time before. With each coop there should be

DUCKS

a sound removable floor, a coop front and a roomy wire-topped run. Have sound coops and runs, not any old box of tricks. Also required are water-pots of suitable size and shape and a tin lid or piece of slate for each coop to act as a table for the ducklings' first few feeds.

We shall also want a piece of galvanized iron sheeting capable of resting on the coop front and covering the run; this for rainy weather; also some fine cut hay or straw with which to cover the coop floor. If the ground is dry we can much better do without the floor during the day; it is better for the ducklings' feet and also the floor will be kept dry. By the way, it is best to put the floor under the sheet of iron near the coop, so that if it does rain you will have a dry floor to put under the coop at night.

Having our three coops, etc., in position with their backs to the prevailing wind, we can now safely proceed to fetch the hens and ducklings. Have you given the hens a really good feed and water? If not, do so. They settle better, brood the ducklings more quickly and do not eat so much of the ducklings' food.

Take the ducklings from under the hen, place them in a suitable warm box, go to a coop, draw out the bar, place the hen in, and give her two ducklings; do not push them under her—let her see them in broad daylight. Then give her the brood, close up the front and come away. Wait 20 minutes, then notice if the hens are brooding the ducklings. All in order? Right, then let us think of food and water.

Procure some water in a jug or bucket, see that it has the chill off, and give it in suitable pots in each coop, just inside the bars. Mix the necessary biscuit meal and mash—allowing some for the hens—and sprinkle a little on the feeding tray, again inside the coops in front of the hen. Take a few ducklings from under the hen, dip their bills in the water, place a few of them on the food tray and sprinkle a little food on their backs (a sure way of teaching them to feed). Break up a little green grass and sprinkle on the water; it moves about and interests the ducklings, teaching them to find the water. Very quickly you will have them all feeding and drinking.

Do not use a coop front; it only darkens the coop. It is better to use the run, which should be of wood, and eight to ten inches in height. By placing one side of it against the coop front, it will keep the ducklings in and the coop will be light.

Very soon the ducklings will be strong and running about. Feed little and often in the run, which should be used in position in front of the coop. As soon as ever possible, work the water pots away from the front of the coop, allowing the hen a drink two or three times each day.

Later, in good weather, a roll of one-inch mesh, 18-inch high netting, can be brought into action, pegged or staked in position, and the runs taken out and kept handy for wet days. As the ducklings grow the pen can be made bigger by unrolling more of the netting. The whole should be moved frequently on to fresh ground. Later the hens can be taken away and returned to their quarters to produce eggs. With each coop I like a wire front, especially during hot spells late in the season. One can use them for very hot nights, making the ducklings safe from rats, etc., yet able to keep cool and get some fresh air.

Make a point of seeing that the coop floors are kept dry and clean. In wet weather it pays to clean the floors every night, and re-bed them with a good sprinkle of cut straw or wheat chaff just before shutting up for the night. The quickest way to kill ducklings is to shut them up, especially when a fortnight or even older, into a coop with little ventilation and wet, dirty bedding. At a month or so of age, the ducklings can be transferred in groups to suitable small houses.

ARTIFICIAL REARING.—Next we come to the question of how to rear successfully a group of 50 or more ducklings by artificial means. It is not really difficult so long as you do not make the fatal mistake of trying to rear in larger groups than 50, or, as a limit, 70. I have had partial successes with as many as 120 in a group, but never again do I propose to try with larger lots than 50 to 60.

BROODING YOUNG DUCKLINGS

Here is a handy type of combination house for accommodating young ducklings. The foster mother on the left is attached to an ordinary 6 by 4 house to make up a very serviceable appliance

Let us go through the different classes of equipment which may be brought into use in rearing ducklings from a day old until they are of sufficient age and strength to require no heat. First, we have the hover type of rearer which has a lamp in the centre and which must of course be in a suitable house, or maybe a brooder house specially for rearing. Then we have the foster mother type, or brooder. This is a brooder which stands in the open and is generally divided into two compartments, one the warm compartment and the other the cool compartment. We then have the battery brooder type, which again must be in a suitable house or building.

All of the above will rear ducklings successfully if properly managed. It is simply a question of choosing the equipment which you consider most likely to suit your purpose. Those who wish to rear only a few ducklings each season can be quite successful with either foster mothers or hovers and suitable houses to go with them.

THE FOSTER MOTHER OR BROODER.—By a foster mother is meant the type of rearer which has two compartments divided by a door. One is heated and the other cool and suitable as a small run and dining-room for the first few days and during very bad weather. This type of rearer is made to stand and work in the open in any sheltered spot or under an open-fronted shed (see photograph). Such brooders may be bought in 100 and 150 sizes, but for practical purposes it is best to have the larger size and then only to use it to rear 60 or 70 ducklings.

In all cases I would advise the purchase of a good six by four by four feet house with a sound loose floor. Make a wooden tunnel of suitable shape and size, the roof of the tunnel to be hinged so that one may easily close the door of the brooder each evening if desired. Use one 25-yard length of 18-inch wide netting of one-inch mesh, and some small stakes. Choose a level piece of ground—if on short sweet green grass, so much the better—place the foster mother so that its windows are facing south-east, put the tunnel connection in position, then the floor and the house. A suitable hole

COMMERCIAL BROODING. Sections of a favoured type of brooder house with sun parlours fitted. It is used for the first month of rearing

should be cut in the end of the house—this and the tunnel should be of good width. Properly arranged, the whole of the equipment should be rat and vermin proof.

Let us now imagine we are getting the whole equipment ready for a batch of 70 ducklings straight from the incubator. The brooder should be properly disinfected and lime-washed. Trim the lamp, and see that the wick is of sufficient length to last the season. Light the lamp, and now for the brooder compartment.

For myself, I prefer to use dry sand as a foundation bedding —1 or 1½ inches spread on the floor, and then a sprinkling of cut hay, cut straw or wheat chaff. Keep handy a box of this material so that you can remove the damp bedding from time to time and add some dry material. Have a thermometer in position. Keep working until you have an even temperature of 85° F., then bed the cool compartment with sand covered with litter. Now pause for a moment and consider the doorway and step between the hot and cool compartments. Can the little ducklings easily return to the warm compartment, especially if a little damp and with full crops? It is best to cut a good turf of sufficient thickness that when it is placed in position it makes a fool-proof run up.

With the hover curtain in good order and the bottom nicely touching the floor, the next thing required is a " stop board," a piece of wood about ¾ by 6 or 7 inches in width and the length of the warm compartment. To this two blocks of wood should be firmly nailed, so that when placed in position in front of the curtain, it will stand firmly on its own. Its purpose is to control the ducklings for the first few days until they learn to return under the hover; without it the youngsters naturally work towards the light of the window and get chilled.

The ducklings should not be moved from the incubator until absolutely dry and strong and active on their feet. Place the ducklings under the hover, then put the stop board in position about three inches from the front of the curtain. Watch the temperature; remembering that you have now added heat by placing in 70 ducklings. For the first three days, run at a temperature of 87° to 90° (it is better to have it at 95° than 80°).

At first, draw back the stop board about nine inches, put in the food tray and the water container, give water and sprinkle food and grit on the food tray. Slip your hand in, work a few ducklings out, and sprinkle food on them. One thing worth doing is to pin up one cut of the curtain; this lets a gleam of light under the hover and causes the ducklings to work towards it and find the food and water. However, if your ducklings are strong you will soon have them drinking and feeding.

For the next three or four days the secret of success is to keep an even temperature, to give food a little at a time but often, and to have the bedding kept dry. Soon the stop board may be taken out, the ducklings having learned to work from heat to food and water, and back again. If the temperature is too low, the ducklings will crowd under the hover to get heat; with the temperature correct, they will be happy and keep working, running in and out from the hover.

In this class of brooder it is as well to leave a little freshly mixed food and clean water the last thing at night so that the ducklings may get a feed and have a drink at daybreak. It must be freshly mixed food or it will go sour in the night.

The cool compartment is used on the fourth or fifth day, after which there is to be no food or water in the hot compartment. Let the youngsters get hungry, then carefully drive them down into the cool room, allow them ten minutes, then drive back any which have not returned to the hover department. When you find that they all return of their own accord you can safely leave the door open. On fine days the lid of the cool compartment may be wide open and the lid of the hot compartment may also be opened a little. Instead of lowering the lamp, try to control the temperature by opening the lid; this means more fresh air and healthier ducklings.

Now for the next move. Provide extra water-pots and open the cool compartment door so that the youngsters may run down through the tunnel and into the house. If desired, pick up a bunch of ducklings and put them in the house. Soon they will be using the whole equipment and we now have them so that they get heat, exercise, fresh air, food and water, and there is no worry about rats, vermin, etc. Give all foods and water in the house, and the further the water is from the heat compartment the better and the drier the bedding will be.

Next, if the weather is good, peg back the house door and peg and stake the netting into position. Soon the ducklings will be in the pen and in fine weather all feeding and watering throughout the day should be done therein. Each evening, however, place foods and water in the house so that the ducklings may feed when they awake.

Throughout the period of rearing, reduce the heat until the ducklings are hardened off and require no heat at all. It is useless to quote temperatures, as these depend on the circumstances, season of the year and how the youngsters grow and progress. Once they are strong and doing without the lamp, they should have the hover lid (to which is attached the curtain) removed. Personally, I like to do this gradually by first turning back the curtain on to the hover lid, then, later removing the whole. Keep the lamp perfectly clean and the wick well trimmed. Don't go the last thing at night and, finding a shortage of oil, fill up the container, trim the lamp, etc., and go to bed, for the next morning you will find the flame up the chimney or the whole brooder burnt up!

HOVERS.—By a hover is meant a type of rearer, usually about two by two feet square (some are round) with sides about 10 by 12 inches high, standing on legs with a curtain all round and with a hinged lid. A lamp and lamp guard stand in the middle.

With a hover we must have a suitable house with a draught-proof floor. I like to nail a piece of three-ply wood on the floor of the house at the place where the hover is to stand; this prevents any risk of floor draughts and the sand and bedding going through any crack or joints in the floor. I then make four stop boards which can be used for the first few days around the hover to control the ducklings and at the same time guard against ground draughts. With the lamp working well, one can proceed in just the same way as when rearing with a foster mother, except that there are a few more difficulties to cope with. Generally speaking, the sooner the youngsters are feeding and watering in an attached wire run, the better.

BATTERY BROODERS.—Here again we have rearing equipment which must be properly housed. Battery rearers are very good in rearing ducklings for the first fortnight or three weeks. Housed in a proper place, which should be run at a fairly even temperature and yet airy and well ventilated, this type of brooder has definite advantages for those who wish to rear large numbers. It cuts out a lot of work and worry when dealing with batches of young ducklings. The youngsters keep remarkably clean, and, running on wire floors, there is little cleaning to be done and no bedding required. For the first few days, clean pieces of old sacking on the wire of the brooder department are necessary, these to be changed from time to time.

As for feeding and watering, any litter splashed and water spilled of course goes through the wire floor, and it requires a few home-made appliances to make it a success. For instance, galvanized trays should be arranged with a spout, and a bucket to catch waste water, which otherwise runs on to the house floor. Peat moss or sawdust should be used on the trays under the wire floors.

As yet, this type of rearer is not much in use for duckling rearing, but in my opinion, after a number of experiments, I feel sure it has come to stay—not for the small rearer perhaps, but for those who have many hundreds of ducklings to rear throughout the year.

Chapter XI

MARKING AND RINGING

THE successful breeder generally has a system of permanently marking his stock, usually when in the baby stage, so that in the future no mistake is made as to breeding, pedigree, age, etc.

Fortunately, waterfowl lend themselves well to permanent distinguishing markings as they have webs between their toes and there is plenty of space for many different toe markings; moreover, these markings may be done when the ducklings are hatched and nicely fluffed out. It is even possible to do it, when absolutely necessary, while the ducklings are wet and just out of the shell. Once marked the mark remains for life.

In any bird there is a left and a right leg. The left leg, commonly termed the " near side " leg is on the left when the bird is going away from you. The " offside " leg is the right leg when the bird is going away from you. The normal duck has three toes and one small back toe on each leg. The back toe is of no interest to us for marking, but between the three toes will be found two webs connecting the two outer toes to the main centre one. On or in these webs we make our permanent markings. Thus we have four separate webs on which to work.

Remember that both the nearside and offside legs have nearside and offside webs, but to make it easier let us call the webs of each foot the outside and the inside webs. Thus we get the nearside outside web; nearside inside web; offside inside web, offside outside web.

It is best to cut a " V " shape in these webs with a small sharp knife, or, better still, a one-sided safety razor blade. If you are right-handed simply take the duckling in your left hand, place the foot on a smooth hard piece of wood and with the blade in your right hand, cut a " V " shape in the desired web or webs. Do not make jagged cuts. One cannot give measurements, it is a matter of judgment, depending on the breed, etc. After trying dozens of ways and ideas I cannot find one half or quarter as good as the one just mentioned. As will be seen from the sketch, many different markings may be made.

In addition, there is a small web on the " outside " of each

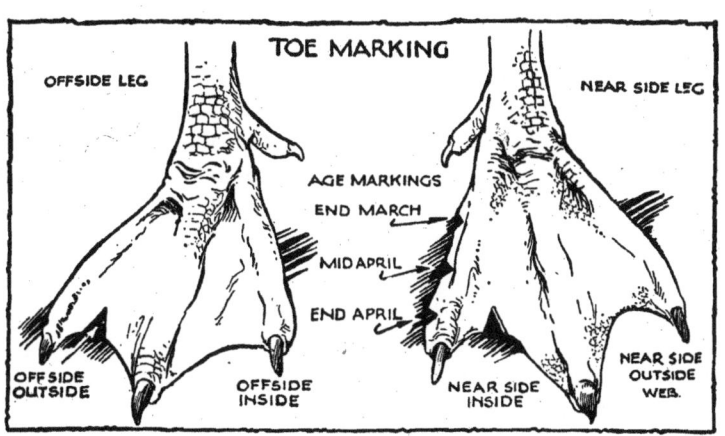

A self-explanatory sketch of toe markings

inside toe and on these one can also cut small marks to give the owner an idea as to date of hatching. This often proves valuable when choosing future breeding stock.

RINGING AND WING-BANDING.—There are two kinds of rings which I have proved from experience to be the best. Remember that toe marking cuts out all question of using different sizes of coloured rings from day of hatching and onwards and with it all the pain which ducklings suffer when the rings are not changed in time for the necessary larger sized ones. Toe-mark as day-olds and the bird carries its pedigree mark for life.

First we have the " Alop " aluminium Conference ring—supplied in the correct sizes for all breeds by Mr. Charles

Allsop, Spencer Street, Birmingham. They are closed rings, oval in shape, each bearing a number and the year. They must be slipped on when the duckling is about six weeks of age and once on they remain during the duck's life. The number and year can be put into Register or Pedigree Book and no mistakes made in the future.

Then we have wing bands. These also can be had with the year and number on and are fixed through the skin in the wing between the butt of the wing and the body. Next there are coloured spiral rings, which may be had in ten or a dozen different colours and of the correct size for all breeds. I say spiral rings because in the case of ducks I have found them to remain on longer and better than flat rings. Different colours may be used for different years—white for one year, blue for another and so on. Use red as danger and keep some red rings by you to slip on any birds which you know to be poor layers or weak in some vital factor.

PINIONING AND WING CLIPPING

If you hatch and rear your own waterfowl it is quite easy, especially when they are day-olds, to make sure your ducks can never fly. There is nothing cruel in this operation so long as it is done when the ducklings are very young, nor is there any cruelty in preventing a bird from flying in the future. The bird that never learns to fly is unable to get out of bounds, and is controllable without high wire netting and all the expense and trouble that go with it. It in no way harms fertility, nor does it make any difference to the value of a bird when killed and dressed for the table. The birds can carry on just as if normal, except that they cannot fly.

To my mind it is far better than having to cut the flight feathers of the birds and by so doing making them scared and wild. The operation is best carried out as follows, as soon as possible after hatching out. Procure a sharp penknife, place the tip of one wing on a smooth block of wood, and remove part of the wing end; really you are removing the part which would later carry the main flight feathers. It is only necessary to do one wing. Properly done, it does not make the bird unsightly as the wing bays or secondaries are left intact and there is no difference to be seen when the bird is going about with wings closed.

Next we come to the question of controlling adult birds which are not pinioned. This means removing parts of certain feathers on one wing.

A bird's wing is made up as follows. In a nearside wing, starting at the left, we have first about ten hard, sharp, long feathers. These are the flight feathers or primaries. The other long and softer textured feathers between the flights and the body are called secondary feathers, or the wing bay or end.

It is the primary feathers in which we are interested. Take a pair of scissors and from the longest primary feathers remove about $2\frac{1}{2}$ to 3 inches. Your bird is not disfigured, yet cannot fly. On no account cut away any of the secondary feathers as this spoils the appearance of the bird and may also cause many unfertile eggs. Another thing is that these secondary feathers afford some protection to the bird's body during cold and bad weather.

Let me repeat that one wing only should be treated in this manner. The effect of clipping is to throw the bird completely off its balance. Clipping both wings might possibly retard flying but it will certainly not stop it.

Chapter XII

DISTINGUISHING THE SEXES

IN the case of fully feathered ducklings, the sexes can easily be distinguished, as the drakes have curl feathers on the top of their tails. In the case of coloured adult ducks, we can go by the rich coloured male markings. Then, above all, we have the fact that the females quack, whereas the male can only make a hoarse hissing sound. Once ducklings are three parts feathered, one can catch each in turn and, by holding by wing, leg or the tail, can quickly pick out the males from the females by their different and easily distinguished call.

When dealing with adult birds of colour, such as Khaki Campbells, Fawn Runners, etc., do not suddenly decide that your drake has turned into a duck! All coloured drakes go into "eclipse" during part of the breeding season, which usually commences in May, and quickly lose their rich colouring, taking on the less conspicuous colour of the ducks. This eclipse lasts until the drake gets his new winter coat, when he again comes into full colour, ready for the winter and his courting displays in the New Year. It is Nature's protection for the well-being of ducks in a wild state—really to make the drake less conspicuous during the rearing season.

SEXING DAY-OLDS.—To begin with, let me say that personally I do not practise the sexing of day-old ducklings. However, some may consider it essential and wish to make an attempt.

It is reasonably easy to distinguish the sexes of young ducklings of a day or so old. In the case of the young males, the copulatory organ can be seen when the cloaca or vent is stretched. In this case, the copulatory organ looks rather like a pinkish root tip. In the cloaca of the female, no such organ can be seen. It is best to perform this examination when the ducklings are strong and fully fluffed out, and just before they go to the rearing equipment—certainly before they have had any food and water.

Take three boxes and work in a very bright light, using one container for the untested ducklings, one for drakelets and another for ducklets or doubtful ones. With the duckling upside down and with its head away from your body, take a firm hold of the tail fluff with the finger and thumb of the left hand. Then, gently but firmly, bend the duckling over the first finger and hold its body there by putting the middle, third and little fingers across it. This, of course, stretches the cloaca lengthways, but great care must be used. Remember you are handling the vital organs of a very young, fragile duckling. Then let the right hand come into action by closing the thumb and first finger and placing them— still pressed firmly together—on the cloaca. Now part them slowly, still keeping a slight but sufficient downward pressure on the cloaca. This will stretch the cloaca transversely and will force out the male organ, if it is there.

Sexing is an operation not easy to explain on paper. By far the best way is to get some experienced person to give you a demonstration. Sexing is of value when dealing with large numbers of laying breed ducklings, saving food and equipment space.

SEX LINKAGE.—Sex linkage in poultry is now very common and much practised throughout the world. It means that by mating different types and coloured males with certain females, we can distinguish the males from the female progeny by the difference in the down colouring. As yet, however, not very much has been done with regard to sex linkage in ducks.

Professor Punnett, of Cambridge, has made some very valuable experiments and he writes: " We are engaged in building up strains of Dark Runners and Dark Campbells. The former crossed with the ordinary Fawn drake will give Dark drakes and Fawn ducks. The latter crossed with a Khaki Campbell drake will give Khaki ducks and Dark drakes. The down difference will enable the breeder to scrap all drakes at hatching ".

From this and other experiments which many breeders are carrying out, it will soon be common to have certain strains capable of giving sex linkage. This would undoubtedly be most useful.

CATCHING AND HANDLING

Ducks are not difficult to handle when once they are caught, but they are in some cases a little difficult to catch. When catching up adult ducks which are in lay great care must be used, as not only can you do much harm to the individual but to others of the flock. Get your eye on the chosen bird, if necessary get assistance, and drive the bird away from the others and catch it in a corner. With the smaller breeds and ornamental ducks it is much better to use a net, in the form of a two-foot diameter ring of stout wire attached to a four-foot handle. The net itself should be made from stout but soft cord. Get the bird away from the others, and then get an assistant to cause the bird to run towards you, down the side of the wire or building ; catch it in the net, give the net a twist and the bird is in a bag and cannot hurt itself or its plumage.

The way to catch a duck is by taking hold of the neck just below the head, then quietly but quickly bring the bird towards your body, slip your free hand over the hand which is holding the bird's neck and slip your thumb and two fingers under where the wings join the body. If you have acted quickly, the movement of bringing the bird towards you will have caused it to open its wings and it is thus ready for the thumb and fingers. Remove your hand from the neck and you have the bird safe for the moment.

Ducks are handled by passing the hand which held the neck under the bird's body and unclasping the hand holding the wings ; when the wings go down, place your hand on top and you have the bird safe and happy.

To take a bird out of a pen or hamper, get hold of the neck

gently but firmly, pull the bird towards you (with some of its weight still on its feet) with breast facing your body; now pass the other hand down the breast and take the bird in your hand. In the case of a small waterfowl a finger may pass between the legs, and your thumb and other fingers form a clasp round the stern and over the wing flights. Never catch a duck by its legs, wing or tail; it is cruel. In a young state, up to three weeks of age, ducklings are easy to handle, but no pressure must be used.

When about six weeks old and when getting their first full feather they are at a difficult stage. They are very tender in flesh and bone, the skin is easily torn and scratched, and the birds may panic and go into heaps, doing untold harm to each other's backs with their sharp claws. After one month and up to say, eleven to twelve weeks it is best, unless absolutely essential, not to catch them or to sort them; they get hot, often causing pneumonia. If you must move them it is best to drive them carefully.

If you have a group of 50 to 60, arrange matters so that you only drive a few at a time into a well-bedded house; it is absolutely fatal to make an attempt with a group in a house or compound.

When ducklings are just coming into first feather, catching or travelling them in hampers or crates by road or rail will often cause a number to go "off their legs," meaning that they hobble about on their hocks or cannot walk at all. This is really nerves. Place them in a sheltered place, with food and water handy and they will quickly recover. Never catch a duck unless it is absolutely necessary, it only undermines its confidence.

Chapter XIII

DISEASES AND AILMENTS

I HEAD this chapter "Diseases and Ailments," as there is a vast difference between the two; in any case, between diseases such as spinal meningitis (or spotted fever), commonly called "staggers," in ducklings, cholera, liver disease, consumption, etc., and ordinary everyday ailments.

Ailments are curable, but in most cases when one gets disease with ducks the very best "cure" is to kill and burn or bury deeply in quicklime. It is much the best way in the long run: do this and you do away with breeding from birds with low powers of resistance to disease.

One may cure a bird or patch it up so that it appears in perfect condition and bloom, yet it is most likely a carrier and will pass on the trouble to its progeny.

It is not worth the risk to spare a weakly duckling; in ninety-nine cases out of a hundred it will never make a good adult bird. It would certainly receive short shrift in a wild state. Nature may be cruel in a way, yet we have only to view wild waterfowl in their natural surroundings to note at once how truly wonderful nature is. None but the very cream lives to procreate the race.

I do not intend to write pages about cures or supposed cures for disease. If you are in trouble, act at once. Take every precaution. Send the bodies away to be examined. POULTRY WORLD has a good post-mortem expert, and there are plenty of reliable laboratories who will give proper and expert advice and opinion. Better still, they will inform you as to the best methods of dealing with that and any future outbreaks.

In short, if you are losing birds or your birds are wrong, kill some of the worst and send their bodies away for examination: it will be money well spent.

DUCKS

Certain minor ailments call for rather different treatment. For instance, if you have toothache, corns or earache, you are in pain, but there is a cure! So let us see what ailments beset ducklings and ducks and consider how best to put them right. This is not difficult, as the duck, if properly bred and reared, suffers from few and only minor ailments.

Colds, i.e., catarrh, which develops into a kind of roup, as in poultry, are seldom to be found in ducks. Occasionally if poorly fed, housed on a wet floor and in draughts, ducks will suffer from frothy eyes; the feathers in feathered birds or down in ducklings is sticky and wet around the eyes, and often frothy white bubbles appear in the corner of the eyes; there may also be a running at the nostrils.

The cure lies in proper feeding, a dry bed (this trouble seldom occurs with adult birds kept in the open) and an ample supply of clean water, in which they can immerse their heads. A mild disinfectant in the water, such as permanganate of potash crystals, sufficient to make the water a bright ruby colour is helpful. A little cod-liver oil in the mash will also help to put matters in order.

This trouble often occurs with ducks which are poorly fed and, say, after the moult. But, given good foods and comfort, the ailment can soon be put right.

"*Slipwing*," or "rough in wing" is definitely a sign of weakness, bad rearing or a check during rearing. It will be found that the first long primary feathers on one or both wings will not lie down properly, and generally stick out. There is no cure other than that the feathers, in a young state, can be tied in position, but is it worth it?

Such birds should never be bred from, for it is most certainly a hereditary weakness. Otherwise sound ducklings can have the rough feathers clipped off short and go into the laying flocks if wished.

Adult birds often become very lame, usually when run in pens without swimming water, and when the grass is worn off the land. Catch the bird, examine it, and you will generally find a hard corn or lump under the foot; the leg is often swollen and is hot to the touch. If rung, immediately remove the ring.

Bed out a small pen or coop, give water and food as usual, and keep applying bread or linseed poultices to soften the lump. Once in a fit state, remove the corn and keep the resulting cavity plugged with lint and ointment. Do not liberate the bird until the cure is complete.

DUCKLING AILMENTS

Scour. This trouble generally occurs early in life, say during the first three weeks. One can usually look for and find a reason: (*a*) Chills and the resulting bowel trouble and scour; (*b*) Floor draughts and insufficient heat under the brooder or hover. In natural rearing the usual cause is allowing the youngsters to get wet through, etc. In such cases the ducklings look sleepy and mope about and the droppings are whitish in colour; (*c*) Badly mixed, stale or sour foods.

There is really no cure as the harm is already done. The great thing is to find the cause, remedy it and do not repeat the error, but profit from the experience.

White Eye. This is undoubtedly a disease and is contagious. It generally occurs in young birds three to eight weeks of age. Symptoms: Wet, watery eyes, the surrounding down or feathers becoming a wet, gluey mass; often a crust will form completely closing the eye and later a white frothy spot occurs.

Causes: (*a*) Insanitary surroundings; dirty water containers; containers too shallow to allow sufficient depth for the immersion of the duck's head; stagnant water in hot weather.

It is best to kill all badly infected birds. However, to avoid having to do this, try to catch the trouble on its first appearance. Isolate, and from time to time immerse the heads in water in which there is a good disinfectant.

Unfortunately white eye is a trouble which occurs during the rearing stages, and unless it is spotted quickly and cured, the duckling seldom grows up into a sound bird. Generally they go off their legs and die.

Pneumonia. Usually caused by incorrect management, too little or too much heat, wet bedding and draughts. If it should prove to be septic pneumonia it is decidedly infectious.

Symptoms: Ducklings very mopy, wheezy, gasping for breath, and very light when handled. The best " remedy " is to kill and burn and to disinfect and remove the cause of the trouble.

Staggers. A most annoying complaint, in most cases incurable, but one that is preventable. It generally occurs in large groups of ducklings at about a fortnight to five weeks of age and is most prevalent during hot weather, with lots of sun and no breeze blowing. It is often put down as sunstroke and, in some cases, is so.

Symptoms: The ducklings stagger about like an old man, falling on their sides, appearing in great pain, and often trying to tie themselves in knots, making their heads touch their backs. In bad cases one can pick up six to eight ducklings at a time in twisted, contorted shapes.

The best way to avoid these cases is to breed from carefully bred matured stock, provide ample shade (natural if possible), a good water supply, and to give plenty of chopped or minced raw onions in the mash. If you have no home-grown supply, Spanish onions are fine. It is said there is no cure, but to my mind onions are a valuable preventive.

Leg Weakness. A difficult trouble to deal with. It occurs sometimes in what appear to be strong, well-grown ducklings. They simply sit about and at first will eat and drink if food and water are placed near them. Gradually, however, they become weaker and die.

In some cases it may be owing to a chill on the liver or kidneys, or even to some liver or kidney disease. If you have many cases in groups of rearing stock, a post-mortem examination is advised.

Breakdowns. I include these immediately after leg weakness, as this trouble must not be mixed up with that ailment. In this case the birds (generally when just in first full feather or with big, heavy adult stock) do not mope or sit about, and generally manage to get along by using their wings.

This trouble often occurs when one has to go into a flock to catch a number of birds, or if they are badly scared after a journey. In a way, it would seem to be a nervous complaint, and it is curable.

Place the patients in a comfortable, quiet place, and with one or two other birds for company. Feed and water as usual. Generally the birds soon recover and become normal.

Sore and Lame Feet. This is not a disease, simply an ailment, and the best policy is to find the cause and remove it. In some cases a thorn or something of the kind may have got into the pad of the foot, or the pad may have become sore through a cut; again, in very dry weather with hard, dry ground, and no swimming water, cracks occur on the under-side of the feet.

Catch the bird, wash the feet and examine them carefully. Should you find a thorn or anything, remove it. If a cut, and it appears sore, clean it well and disinfect with iodine. Keep up on a soft dry bed until cured. If caused by dry weather, provide some " paddling " water.

There are other duckling ailments which occur from time to time. Many are unaccountable, such as when the birds in the baby stage will not eat. They are usually in-bred or from immature stock. Other troubles may be due to bad management, improper foods, etc.

Chapter XIV

KILLING AND PLUCKING

DUCKLINGS come into first full feather at eleven to twelve weeks of age, the exact age depending on the breed, rearing and season. Muscovies are the exception to this rule as they are not in full feather until sixteen weeks. At eleven to twelve weeks, ducklings will pluck out fairly free from stubs.

It is, of course, up to the owner to use his judgment as to when the birds are ready for killing, both as regards feathering and condition. If the ducklings are fat and the season early it often pays to kill when the birds are nine to ten weeks old, or when they have their wing feathers half to three-quarters grown. Once ducklings are in feather they should be killed, plucked and sold.

If you miss the first feather stage, when dealing with table ducklings, it means keeping them on until they moult and come into second feather. Immediately they get their first feathers they commence to moult and grow their second lot of body feathers; the wing feathers are not moulted, only the body feathers. It will therefore be seen that the secret of success with all table breed ducklings (or surplus breed drakes) is to have them fat and in prime condition ready to kill at ten to twelve weeks. If by poor management and feeding they are not ready at this stage, you have missed your main profit and quick turn-over.

To make a success with table duckling production, one must own the right type of bird, and keep the ducklings on the move in growth and flesh all the time, from the first feed until they are ready to kill. They must never be allowed to get lean, or run as " stores " with the idea of giving a last fortnight's special fattening. They must be kept growing and fattening the whole time.

KILLING.—When ducklings are fit for killing, the usual mash food is given late on the afternoon or evening of the day before they are to be killed. Handle them with care when catching and place them in cool, airy crates.

The best method of killing is to take the birds by the legs with one hand and then dislocate their necks with the other. Hang the bodies up by the legs so as to allow the blood to run into the neck. Remove all large wing and tail feathers; these are of little value and generally go on to the manure-heap. Allow about five minutes to drain, then commence to pluck; the body feathers should be saved for there is always a sale for them.

When plucked, birds should have any stubs removed and this is best done with an old blunt pocket-knife. Clean the legs and feet, then bend them over so that the bottom of each foot rests against the back of the duckling. Turn in the wings, and put in a cool place on a cold surface.

On no account must any duckling be packed for dispatch until thoroughly cooled. They are best packed in butter paper and on layers of clean wheat straw in airy hampers or flats when dealing with small numbers; with large and regular numbers it pays to pack in proper boxes, with the birds graded to size and quality. There is definitely a fine and regular demand for prime quality table ducklings during the months of February to early July and, of course, at Christmas.

CHAPTER XV

ORNAMENTAL WATERFOWL

THIS chapter could really be called "Ornamental Wildfowl," as the ducks mentioned are naturally wild waterfowl kept in captivity and bred from. Most of the species mentioned are very ornamental in colour and outline.

Birds bred in captivity can now be had in many species and such will generally breed in captivity if the surroundings are reasonably good. This cannot be said of wild-caught birds, which will hardly ever breed, although the males prove useful in introducing fresh blood, as they will generally mate with the hand-reared females.

Most of the commoner Ornamental species are quite easy to keep. They are very hardy, inexpensive to feed, and they will stand the rigours of an English winter unhoused and with only some natural shelters in the form of evergreen shrubs. But they hate wind and must have shelter from it.

With a small garden anyone could keep a few pairs in a wired-in enclosure with a small cement pool in the centre and the pen planted with ornamental shrubs and trees to form shelter. A very small pen would do for a pair of such as Carolinas or Mandarins or different species of the Tiny Teal family such as Common Teal or Gargany Teal.

The two first mentioned are certainly the most beautiful and gaily coloured of all waterfowl. The Carolina is a native of North America, the Mandarin comes from China—both mate in pairs and are very faithful to their mates. In a wild state, or if left unpinioned in suitable surroundings, they make their nests in holes in a high tree, so that when kept in captivity they must be given a suitable nesting site in the form of a 12- to 15-inch square box, best made in the form of a dog kennel with a four-inch square entrance. The box is best placed about three to four feet from the ground up an old tree trunk or in a thick bush.

I once acquired a number of old tree trunks about one foot in diameter, dug holes and placed them firmly in the ground at an angle with a good prop under the top end; then, working with natural wood, I nailed a framework on to the tree trunk to hold the boxes and arranged an easy run-up of latticework with a wide entrance step to the box. I cut a thin grass turf to place in the box to form a base and threw in some rough dry grass. The rest was left to the birds.

In some cases it was possible to get very rough barked stumps of trees which required little ladder work other than a good landing in front of the entrance. Also, it improves matters if the box is roughly covered with branches from conifers; it makes things more natural and private.

The Mandarin duck lays from the middle of April, white eggs about the size of a large bantam egg; generally seven to eight eggs in a clutch. The incubation period is 30 days.

The Carolina lays earlier than the Mandarin; much the same type of eggs only a little smaller. Incubation period, 30 to 31 days.

In both cases it will be found best to take the first clutches of eggs and set them under bantam hens, but some with ideal pens and situation may care to allow the duck to hatch and rear her own young. It is certainly more interesting and very fascinating to watch.

A pair of either of these small and beautiful waterfowl will live and breed on the floor of quite a small aviary (with suitable nesting box in proper position) and require only a small cement basin, one about six inches deep and 18 inches in diameter; as a matter of fact it is best to be small so that the water can be easily changed. The ideal pool is one with a plug and runway. Another point to watch is to have the sides of the pool well above water level, so that the birds can easily get in and out but do not make the pen messy and muddy.

DUCKS

For food they require little other than clean sound wheat for most of the year; this can be fed in the pool so as not to harm any of the other birds in the aviary. In the New Year the ducks can have a small feed of soaked or scalded biscuit meal which has some meat meal in it, dried off to a crumbly state with some very good first feed duckling mash—this often has a percentage of cod-liver oil in it.

CAROLINA DRAKE
One of the most beautiful and distinctive types of duck entering the ornamental class

Throughout the summer the young ducks and adults which are not on natural waters receive a ration of " duck weed " or chopped lettuce two or three times each week—placed in the water in the cement pools. During the moulting season the adults receive a mash ration to help them to acquire their new apparel.

Other species which do quite well when kept in pens with cement pools are Pintail, Gargany Teal, Common Teal, Mallards, Black East Indians, White Decoys, Brown Decoys, Hooked Billed Ducks of Holland, Widgeon (these are grazers and will eat lots of short green grass when they can get it), Bahama Pintail, Gadwell, and others which are surface feeders.

Then we have the fascinating Small Diving Ducks, very interesting to keep and to watch. They seldom breed, unless kept under ideal conditions with natural water and with reed-beds, etc. They also require good feeding, with some animal matter in the ration such as meat meal or white-fish meal. In a natural state they work very hard and must consume lots of water snails, etc. To watch them at work on natural waters is most interesting—they are generally followed by surface feeders, who eat what the divers cause to rise to the top of the water.

The following varieties are also hardy and well worth while. I have even known them to be kept in a 20-yards square grass pen with a pool about six feet by four feet and two feet deep. The Tufted, often called the Magpie Diver. The drake has a black head glossed with purple and carries a good big crest; the eyes are bright golden-yellow, extra bright during the mating season, and big for the size of the bird; the neck, breast, back and tail are black, the flanks and underparts pure Chinese white—a really cobby little bird with a very quaint gait when walking. Generally they do not lay until late in May. The incubation period is 23 days. The nests are built in clumps of reeds in shallow water.

Then we have the Common Pochard, built on much the same lines as the Tufted but a little larger and rather cobbier in body. The drake in full colour is quite a pleasing sight with his bright red eyes, head and neck coloured a deep chestnut, and the body light grey and finely pencilled. Eggs are generally laid in April—about ten is the average number per nest, which is built in clumps of reeds in shallow water. The incubation period is 25 days.

There are lots of other species of diving ducks such as

American Pochard, Red Crested Pochard, White-eyed Pochard, Golden-eye, etc., all well worth keeping in a collection if one has natural waters to put them on.

If it is wished to keep just a few pairs of these diving ducks and you are making an artificial pool, have it at least two feet deep and easy of access, for the birds are not very agile when pinioned, although they are good and very fast fliers in the wild state.

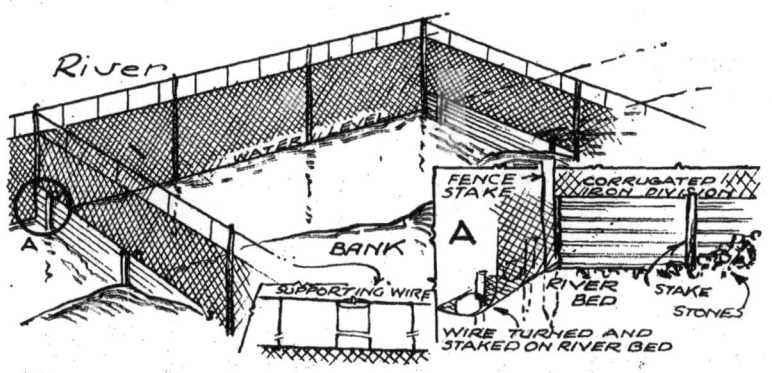

A handy swimming pool which is as suitable for a number of ornamental ducks as it is for a small flock of breeders

With a good sized pool one can get endless amusement and interest watching the antics and diving aquatics of these birds. They are very tame and friendly.

Other varieties quite interesting to keep are Common Shellduck and Ruddy Shellduck, but they require a fair amount of animal food to keep them in good health and to get them in breeding condition. They also eat plenty of grass.

They are really beautiful but inclined to bully smaller birds in the breeding season. Personally, I question if they are really ducks—they are more like very small geese in some of their habits. The incubation period is 28 days for both varieties. In a wild state they lay their eggs in any old hole —in a rabbit burrow or in a crevice between crags or rocks. When kept in captivity it is necessary to provide artificial holes—best done with drain pipes with a box at the end. Have the box top covered with turves so that one can examine the nest when necessary.

The young are quite hardy and not difficult to rear so long as they are given some animal food, chopped worms, chopped green food and duck weed with snails and insect life among it.

When rearing such as Diving Ducks, Shellduck, and in fact, most of wild duck tribes, it is found that they rear quite well for the first week or ten days and then, unless they receive duck weed and special foods they gradually die off. If you have natural water, especially a small stream or shallow river, the best plan is to arrange the coops on the bank and with wire netting and stakes to make a pool in the natural water.

The Carolinas and Mandarins are not too easy to rear in coops with hens and are often a little difficult to start feeding. However, one way I have found efficient is to place one or two small Brown Decoy ducklings or Mallards with them, or

ORNAMENTAL WATERFOWL
Here is one of the cement pools described by Mr. Appleyard in an earlier chapter

one or two of their own species which are already feeding well.

This chapter is written with the idea of interesting the reader who may keep a pair or two in a garden pen with a small pool, or in an aviary used by such as budgerigars. Or the reader may be fortunate enough to have a natural stream or an old pond, or even a small lake. All these can be made beautiful and interesting if properly stocked and arranged.

A small pond, if it and some of the surrounding land is properly fenced and wired in, with, if possible, a small island in the centre, can be made very beautiful with ducks. Even an artificial island is easy to make and stake securely in position —just a rough wooden punt filled with soil and turf and a bit of bush work so that it just rides nicely above the water and the ducks can easily get on and off.

The photograph on the previous page depicts an oblong pool and a small "Saucer pool" in my garden, made by myself; a number of pairs do quite well in it, including two pairs of Carolinas and a pair of Mandarins as well as Diving Ducks. From these I get a lot of amusement. They know me, look forward to my visits and will feed from my hand with confidence, although quite a few are hatched from wild-gathered eggs. They are hardy, inexpensive to keep, and they live in the open winter and summer. For a house they have clumps of evergreen shrubs and small conifers.

When constructing a pond of this description do not overlook the necessity for regular draining. Failing a small pump an excellent plan is to fit drain plugs at each corner, first taking the precaution to sink a fair depth of rubble at these places before making the bed of the pond.

Chapter XVI

SHOWING

ALL who keep ducks do not look upon them purely as layers of eggs or as table meat. Many like to exhibit them, thereby deriving an added interest and pleasure and at the same time advertising ducks generally, letting the public see you have good birds and most likely making other enthusiasts.

There is no difficulty in exhibiting ducks. Roughly, this is the procedure. First of all, send a postcard to the secretary of the Show asking for a schedule, and when it arrives, look through the classes. It may state, " Class 19, Aylesbury, Pekin or Rouen, Either Sex," and this means you can enter as many males or females as you wish in these three breeds. Other classes might read : " Indian Runner or Khaki Campbell, Either Sex " ; " Duck or Drake, A.O.V.," meaning any other variety, or that you may enter any breed with the exception of those already named.

In the case of " Indian Runner or Khaki Campbell " you would be quite in order to enter any colour of Indian Runner ; on the other hand, if the class read " White Runner or Khaki Campbell," you would be wrong if you entered a Fawn or a Fawn and White Runner, which in this case would have to be entered in the A.O.V. class.

With the schedule will be an entry form, which should be read carefully ; also the rules concerning entries. On the entry form you will find columns for class, description of entry, entrance fee, and sale price. In the first column you put the number of the class in which you wish to enter a bird ; in the next the description, such as " White Indian Runner drake " (in some cases you have to declare the year of hatching) ; then the entrance fee ; and, lastly, the price at which you would be willing to sell the bird, the figure to include the hamper. If you do not wish to sell, you can put on a prohibitive price or mark the form " N.F.S."—which means not for sale. In any class you can make as many entries as you wish, unless there is a rule against it in the schedule.

Next you will need the right type of light show hamper; a lot of money can be saved in rail charges by owning the proper kind of light wicker canvas-lined hamper, and your birds will travel more comfortably and arrive in better condition. The hamper should be well bedded out, preferably with hard, clean wheat-straw with a covering of loose chaff.

The secretary of the show will have sent you the necessary labels, on each of which will be a Class Number and Pen Number; look up the schedule and see that you get the right label on the right hamper. On the labels you will find a loose tear-off piece containing the address of the show; raise this and you will find a space to be filled in with your own name, address and railway station, for the return of the birds.

See that the birds catch a train which will get them to the show in good time. You pay full rate for the outward journey and half-rate for the return; they must, of course, travel by passenger train.

PREPARING BIRDS FOR SHOW.—To obtain success when exhibiting ducks the birds must be prepared and shown in perfect condition and bloom. Always remember that the good bird shown out of condition and wild will generally be beaten by even the moderate bird shown in perfect trim and trained to the minute. A judge has not the time to spend in trying to make a bird " show " when it is wild, scared and skulking in the corner of a pen. Moreover, such birds are no advertisement for their owners or their breed. Also, it is both cruel and unfair to pick up a bird off the range, place it into a hamper and send it to a show.

Arrange a few training pens in some building, preferably in a place where people are frequently passing the pens, and have the bottom of the pens about three feet from the ground; it is really best to purchase a proper set of three, four or more pens, for they prove very useful for sick or maimed birds, periods of isolation, etc.

Having arranged the training pens in a suitable place, the next thing is to get them occupied. Here let me say that it is next to impossible to make any sort of a job of washing a white or light coloured duck; they are best left to wash themselves on natural waters or in cement pools.

In many cases, where only a few birds are kept, it is quite possible to pick up the best birds at the evening feed. The best way of all is to have a small house near the washing pool and to keep separate the birds which you propose showing. Have the house well bedded out with dry straw, drive the birds from the water into the house, close the door and leave them alone for a quarter of an hour. Now catch up the birds required, place them

Show Type Khaki Campbell Drake

DUCKS

in the show hampers, proceed to the penning room and put the ducks in the pens; if wished, place two birds in a pen for company.

Never keep a white duck penned for longer than two days; they get stained and as they cannot wash the feathering gets very soiled and takes ages to become white again.

Cut a stick about two feet long, or use a judging stick, and move the birds about with it. Talk to them, visit them as much as possible while they are penned, feed and water them, but give both on the sparing side; by being a little hungry they will take an added interest in your visits. From time to time give them a few worms and similar tit-bits; a few such dainties, kept handy in a tin in damp grass or moss, will work wonders.

For the last two or three days before the show the birds should be on the range. On the day they are to go on rail and about one hour before it is necessary to catch them,

Show Type Khaki Campbell Duck

give a feed of grain. Drive the birds into their house and leave them for a time until dry; if you drive ducks into a house and proceed to catch them immediately, you will find they have a certain amount of moisture about them, especially their legs and feet, and this often spoils the bloom of a white bird as they are inclined to trample on each other unless very tame and some care is used.

Having caught the birds, wash and wipe their bills, legs and feet with a suspicion of vaseline and wipe well with a clean, dry cloth. Remove any private rings. If the show secretary has sent numbered rings, affix them on the specified leg, upside down so that the number can be read when the bird is in the judge's or steward's hands. Lastly, into the hampers and away to the show.

If the show is near, try to visit it. When you get there, get to know other exhibitors and breeders. If you do not win, don't grumble; on the other hand, ask questions and make an effort to get to know why the red-carded bird was a winner and yours only Reserve or Highly Commended. If you can get hold of the right party, he will explain why the birds were so placed. You may possibly have got a much better bird at home! Get fixed in your mind's eye what is wanted, then go all out to breed and exhibit such. Remember it costs not one halfpenny more to hatch, rear and feed a good bird than it does to do the same with a poor or mongrel one.

Chapter XVII

NOTES FOR NOVICES

THE period of incubation with duck eggs varies according to the amount of natural or artificial heat given to the eggs over a period. Light breeds usually take 26 to 28 days; heavy breeds 27 to 29 days. The period is generally shorter in incubators than under hens or ducks.

The Muscovy duck is an exception, as the incubation period is 35 days. It will mate with big breed ordinary ducks but the progeny are sterile and will not reproduce. Incubation period then is 33 days.

On an average, ducklings are in first full feather at 11 to 12 weeks. That is to say, they are nearly free from " stubs " if killed and plucked at this age. Ducklings will pluck out fairly free from stubs when the long flight feathers are three parts grown but are best when first full grown.

To miss killing at 10 to 12 weeks of age means that the ducklings start to moult their first body feathers and a second lot begin to grow and the birds must be run on. A duckling does not moult its wing feathers (primaries and secondaries) at 12 weeks, only the body feathers are moulted.

To distinguish the sex of ducklings : Young ducks quack, young drakes only make a hissing noise. When in full adult plumage the drake has a curl on the tail, generally three curly feathers ; the duck has no curl.

In light breeds a drake can generally mate successfully with one, two, three, four, five or six ducks ; two drakes with ten to fourteen ducks ; four with thirty ducks and so on. In heavy breeds much depends on the type of drake used. In the early season a drake with two, three or four ducks ; later, in milder weather, five ducks. With large exhibition types it is best to mate a drake and two ducks. With utility heavy breeds a drake will mate with four or five ducks ;

two drakes with eight to ten ducks; three drakes, fifteen to eighteen ducks.

All coloured drakes, such as Mallards, Khakis, Fawns, etc., have an annual "eclipse" or suffer a partial moult during the latter part of May. This is Nature's way of making them less conspicuous during the rearing season. In the autumn the drakes moult out and then acquire their gay plumage ready for mating in the New Year.

Duck feathers, if stored carefully, are saleable and will fetch a good price. The quills and hard feathers from the tail and wings are of no value except for manure.

A useful rule is to reckon $5\frac{1}{2}$ months as the average time taken by a well-bred laying ducklet of a light breed to come into production. Heavy breeds, six to seven months. In both cases much depends on the season, and when hatched. The useful life of a laying duck is generally three seasons. As a breeder it can be used for many years.

If you want results with young or adult waterfowl, remember to have a dry floor and plenty of top ventilation. Lots of food with elaborate housing will not alone give successful results. Good food, properly prepared and fed, plus quietness, care and cleanliness are necessary. Quiet and gentleness will give that confidence and tameness conducive to success.

Use only the best. The best costs no more to feed, or house than second-rate stock. It is better to rear a few birds really well than attempt numbers beyond your capacity.

Know your birds, learn stock sense, use your eyes all the time. To know each individual bird, its ways and characteristics is invaluable to success in breeding. If you are interested in any particular breed, go to a few big shows, see your breed and talk to other breeders and exhibitors, or, better still, exhibit your best birds and see how they compare with others. Do not ask advice from too many people. Choose your adviser and then follow carefully that advice in detail.

A good broody hen is best for rearing ducklings. She will brood them adequately, whereas the mother duck when cooped, makes a wet mess of the floor and does not brood her ducklings sufficiently. Cold weather, damp and chills cause death; use a hen whenever possible.

Remember when choosing your breeding stock that size is an inherited factor. You may improve size by good rearing and proper foods, but remember that to get size you must breed for it, therefore choose good birds which have body size and which handle well.

It is said that doctors bury their mistakes, solicitors send in a bill and that live stock breeders often look upon their mistakes for a lifetime. Remember this and when mating your pens do clearly understand that what goes into the mating will come out, often many generations later.

The beautiful bird with just a suspicion of " wry tail," use her and for few or more generations you will have birds coming with this weakness. It is the same with all faults in live stock, much more so to my findings than good points. The perfect bird is not always capable of passing on his or her good features and it is here that good pedigree and breeding repays. Health, stamina, liveability, hatchability and rearability can and must be bred into a strain if the strain is to be of value. Never try to rear a sickly youngster, kill it, spend the extra time you would have expended on it, on the other youngsters. Never cure a sick bird and then use it in your breeding pen, it is not worth while in the long run.

Nature may be cruel, in wild life certainly cruel, yet there is left the very choicest of stock to carry on the race. In bird life only those survive which have stamina, brains and a fitness sufficiently worthy to carry on and produce perfect progeny. So do remember cut out the " poor little thing " business. Kill it painlessly and give the food it would have eaten to a better little one !

Lastly, never make your birds into " greenhouse " stock. You must breed for stamina and health, it is the only way.

GLOSSARY OF DUCK-KEEPING TERMS

ADDLED.—Term used to denote a bad egg when it has been incubated.

BARS.—The bars of colour in the wings of coloured drakes or ducks such as Rouens; the two white bars in a Rouen's wing divided by the blue.

BEAN.—The tip of a duck's top mandible (beak).

BILL.—The beak.

BROWS.—Overhanging or heavy eyebrows often seen in a coarse Aylesbury, Rouen or Pekin.

BURST YOLK.—When an egg is placed in the incubator and tested the yolk is burst and of no further use.

CANDLING.—Testing eggs before a light to examine contents

CARRIAGE.—The manner in which a bird carries itself. Low, horizontal body carriage is an attribute in a Rouen or an Aylesbury; an Indian Runner should have an erect carriage.

CHAIN ARMOUR.—Used by fanciers in connection with Rouen drakes, meaning that the feathers in the claret colour are pencilled or laced with grey—a bad fault in an exhibition Rouen.

CLARET.—The breast colour in a Rouen drake—also, in a lesser degree, in the Mallard drake.

CLEAR.—Used when talking of eggs which are incubated and show no life when tested. Clear, like a fresh laid egg.

CREST.—Generally denoting a small compact, globular feather on the head of a bird.

DEAD GERM.—Used to denote an egg which was fertile but in which the germ, which is attached to the yolk, dies.

DISHED BILL.—A term used to denote a duck which has a fault in the bill. Taking a line from the tip of the bill to where it fits into the skull, the line, when viewed from the side, is concave or hollow. The tip of the bill is turned upwards which makes the top line appear dished. This is a serious fault in such breeds as Runners and Aylesburys on the show bench.

DUCKS

DRAKELET.—A current season male bird.

DUCKLET.—A current season female.

ECLIPSE.—A partial moult of the coloured drakes in early summer, generally the latter end of May, when the males cast their gay plumage and take on the colour of the ducks.

EYE STRIPE OR EYE STREAK.—Meaning dark or light markings on the head of a duck, generally running from the base of the bill to above and below the eyes and towards the back of the skull.

INFERTILE.—In the case of a bird, meaning one which will not reproduce. In the case of an egg, one which shows no germ.

KEEL.—Loose flesh, in the shape of a boat's keel, covered with feathers. When viewing a good Rouen or Aylesbury from the front, the breast comes out in a wedge shape and from the side is well forward; this is called the keel. A Pekin or an Orpington should have a well-rounded, "apple-shaped" chest, free from keel.

MALLARD.—Another name for the wild duck—often used in the expression "a Mallard marked duck," meaning markings and colorations in domesticated ducks similar to those in the Mallard duck or drake.

PINION.—The last small joint in a bird's wing—the part to which the long flight feathers are attached. This pinion is removed if it is wished to make a bird unable to fly—but it should only be done during the first few days of life.

PRIMARIES.—The first ten large pointed feathers in the wing. When the wing is opened these feathers are furthest away from the body.

RING.—The white ring which divides the neck and breast colours in such breeds as Rouens and Mallards.

ROACH BACK.—Viewing the bird from the side the whole line of back is convex. By handling such birds it will be found that the back is arched and the backbone out of line. In no circumstances should such birds be bred from. Roach back and wry tail are often found in the same bird.

ROMAN NOSED.—A term used to denote the exact opposite to dished bill, the top line being convex. This feature is often useful in certain breeds to breed stronger and more wedge-shaped bills.

SAPPY.—A term used to describe a white bird which is poor in colour, showing a yellow tinge.

SECONDARIES.—The flight feathers on the wing between the primaries and the body. In the case of coloured breeds they are the feathers which carry the colour.

STUBS.—A term used by pluckers to indicate that a plucked bird has lots of small young feathers just coming through the skin.

SWAN NECK.—The best way to understand this term is to note the neck of a swan. In a duck the back line of the neck has a definite "hump" when viewed from the side. When the back line goes right up to the base of the skull in one sweep it shows a convex curve—a fault, and one which is difficult to eradicate if bred into a strain.

TWISTED WINGS.—Often termed "rough in wing" or "slipped wing" a decided deformity and one which should debar birds from the breeding pen. The last bony joint or "pinion" on which the long flight feathers grow is weak and the long feathers stick out from the body, or are rough and never keep properly in place when the bird is in repose.

WEB.—In the case of waterfowl the flat skin between the three front toes. It also denotes the blade of a-feather. Also used to denote the part in a wing of a bird where wing bands may be attached.

WRY TAIL.—If, when looking at a bird from above, the tail is carried on one side, out of line or twisted, it is termed a wry tail, a bad fault which generally denotes a weakness in constitution. Such birds should never be used as breeders.

DUCKS or HENS

Which Are the Better Egg-Machines?

WE all know the story of the man who kept his hens in the cellar, and who, when the water-pipe burst and drowned them, philosophically remarked: "*I wish I had kept ducks.*"

That's as near duck keeping as many poultry keepers get. But with strains of ducks averaging 250 eggs per annum — when most of us are content to get 200 eggs from our hens — and with Karswood Poultry Spice to help ducks live up to their reputation as high-output "egg-machines" — it certainly behoves hen-men to consider if they have not room for a few ducks — **as a trial.**

Four Daily Layers

Albert Strong writes from " Somewhere at Sea "

"I have found Karswood Poultry Spice has made it possible for me to keep my 4 Khaki Campbell ducks on the ration of Balancer Meal for one bird. I am writing this on the high seas while serving in the Royal Navy.

My 4 ducks recently laid 63 days consecutively — laying 4 eggs a day. A total of 252 eggs! This remarkable feat I attribute to Karswood Poultry Spice, as my Balancer ration is for one bird only."

Another duck keeper, Mr. J. B. Greenwood of Teddington, Middlesex, reports:—

"We have four Khaki Campbell ducks — three adults and one about 4 months. The adult ducks have laid consistently well. Two of them laid over 100 eggs each without a break between January and April, and we are still getting 15 to 16 eggs per week from the three. We have used Karswood Spice ever since we started."

The Natural Egg-Producer

It should be borne in mind that Karswood Poultry Spice is **not** forcing in its effect. It contains **NO** forcing ingredients, but it **does** contain ground insects and other tonic ingredients which keep the bloodstream rich and pure so that it is always full of the nutriment which the ovary must have to ripen the tiny egg "seeds" clinging to it, and to turn them into eggs.

Note the Economy

To duck-owners — new and old — we say: Unless you have tried Karswood Poultry Spice in the mash, you will never really know how many eggs your birds are capable of laying. One farthing's worth of Karswood Poultry Spice is sufficient for 10 ducks for one day. So you will agree that you can afford to make the trial! Karswood Poultry Spice is sold by all Poultry Food Dealers in packets 2½d., 7½d., 1/3, 3½-lb. bag 4/2, and larger bulk sizes.

KARSWOOD POULTRY SPICE

ADVERTISEMENTS

DAY-OLD
AYLESBURY also KHAKI CAMPBELL

Ducklings from March to October. Let me know your requirements and I will supply at a reasonable price and give satisfaction.

WILKINSON
PARROCK FARM, HEBDEN BRIDGE Tel. 195

HERIZ SMITH
KESSINGLAND BEACH

WHITE CAMPBELLS (PEDIGREE)
APPLEYARD'S STRAIN

DUCKS!! DUCKS! DUCKLINGS

We don't care if it's two for Granny Jones or 5,000 blood-tested 6-week Guaranteed Females for Poultry Farmers who realise that duck farming with the right stock is profitable. Also Breeding Pens of Magnificent AYLESBURIES, KHAKI-CAMPBELLS, MUSCOVIES
at rock-bottom prices

See Adverts in " Poultry World "

DUCKERIES
WORCESTER PARK, SURREY.
Derwent - 4749

PEDIGREE
KHAKI CAMPBELL DUCKS

1943 hatched.   280-300 egg strain

FROM **35/-** EACH

Also **UNRELATED**

KHAKI CAMPBELL DRAKES

FROM **35/-** EACH

Breeding pens and stock birds always for sale

Can be seen any time, any day

All on 7 days' approval

10/- deposit on crates, returnable.

DEEPDENE PEDIGREE FARM
(C. H. WAYMOUTH)

BROAD LANE, HAMPTON, MIDDLESEX

PHONE : MOLESEY 2171

LIMESTONE GRIT
FOR
DUCKS, GEESE, TURKEYS, BANTAMS, POULTRY OF ALL AGES

Hard mountain Limestone—White, fine and dry—No Dust—graded and clean—99·20% Carbonate of Lime and 100% British—Guaranteed analysis—In new bags

> Mr. REGINALD APPLEYARD writes in his book on "Geese," published by "Poultry World":—
> "One thing is certain, soft-shelled goose eggs are most useless and annoying things, and are to be avoided. I have found Limestone Grit and Flour most useful."

"LIMESTONE" GRIT (Adult or Chick Size)
CARRIAGE PAID nearest Goods Station, England and Wales
1 cwt. 6/6, 5 cwt. 29/-, 10 cwt. 55/6. Larger Quantities quoted for.
"LIMESTONE" FLOUR (for Mashes and Mineral Mixtures)
SAME PRICE AS GRIT

J. THORNHILL, Great Longstone, Derbyshire, 45

MOPING unprofitable DUCKS
QUICKLY CURED
WITH
JOHNSON'S REVIVING TONIC
(REG. TRADE MARK)

ENSURES HEALTHY PAYING BIRDS

Renowned for over 35 years

Bottles 8½d., 1/4, 5/8 from Corn Stores everywhere

JOHNSON BROS., Poultry Medicine Manufacturers, WEST BROMWICH

HAMER'S QUALITY DAY-OLD DUCKLINGS

The Firm with a Host of Satisfied Customers

Aylesbury - Khaki Campbell

Write for List

GEORGE HAMER, Moss Hill Farm, Bradshaw, Bolton, Lancs.

Phone: Eagley 280.   Est. 34 years

GOODCHILD BROS.

BLACK CORNER, Nr. CRAWLEY, SUSSEX

RABBIT SPECIALISTS OF 26 YEARS' STANDING

Have 5,000 rabbits from which you can make your choice. RABBITS FOR BREEDING, Rabbits for exhibition purposes, fur, wo― S STARTED.
Requirements of small breeders attended
EVERY RABBIT SUPPLIED WI' ACTION
Wild and tame ra

ENGLAND'S LE **3ITRY**

BRITISH DUCK KEEPERS' ASSOCIATION

INCORPORATED AND ALLIED CLUBS:
BUFF ORPINGTON D.C.; INDIAN RUNNER D.C.; UTILITY I.R.D.C.; KHAKI CAMPBELL D.C. (including Whites).

The body to which all Waterfowl Keepers should belong—so as to ensure proper consideration of post-war problems

War-time Subscriptions (for duration of war or 12 months—whichever is the longer):—

ASSOCIATE—under 30 head of stock—5/-
MEMBER, 10/-   FELLOW, 21/-

Apply to Hon. Secretary:

ST. ANTHONY'S, SWANLEY, KENT

PETER BROOKE
KHAKI - CAMPBELLS

A keen interest in Khaki-Campbell Ducks, coupled with years of selective breeding for the right stock, backed by the necessary equipment for rearing better and still better Khakis . . . . Peter Brooke Khakis equal the best.

GROWING AND LAYING DUCKS
STOCK DUCKS AND DRAKES
Usually available.

PETER BROOKE LTD.
DEMONSTRATION P. FARM
OLD COULSDON, SURREY
(UPLANDS 2457)

POULTRY KEEPERS!
IT PAYS TO INSURE

Stock is not covered by comprehensive household policies AGAINST

1. Theft.
2. Death by the notifiable diseases.
3. Seizure by Fox, Dog or Vermin
4. Storm, flood or tempest

ANNUAL PREMIUM 6d. per bird.   Minimum premium 4/-
SPECIAL COLLECTIVE POLICIES FOR CLUBS AT REDUCED RATES.

Send for prospectus (1d. stamp—Govt. order) at once to:—

HERTFORD INSURANCE CO. LTD.

Reg. Office : 36 St. Martins Lane, W.C.2
Emergency Address :
Chessway, Loudwater Lane, Rickmansworth Herts.   Ricky 3718

Mrs. K. F. STICKLAND, U.D.C.

BREEDER OF THE FINEST UTILITY

Buff Orpington

DUCKS

EGGS AND STOCK DRAKES

USUALLY FOR SALE

The Old Downs
LONGFIELD, Kent

Phone. Longfield 2109
Station: - Fawkham

Success with
KHAKI-CAMPBELLS

depends on having a healthy, heavy-laying strain. **Aylsham Khaki-Campbells** produce an abundance of eggs and are a profitable investment.

Send for prices of hatching eggs and ducklings to :—

A. C. S. BOWMAN
Manor House, Aylsham
Norfolk.
Accredited Breeding Station

DOWLOW
PURE LIMESTONE GRIT

was first in 1900 and still leads the way

1 cwt. **6/6**.   2 cwt. **12/-**.   5 cwt. **29/-**

New Sacks. Carriage paid nearest Goods Station in England and Wales

PURE LIMESTONE FLOUR

Same price as Grit

DOWLOW LIME & STONE CO. LTD.
Dept. C, BUXTON, DERBYSHIRE

FOR ILL-HEALTH IN DUCKS

As a powerful germicide, Iglodine Antiseptic is an excellent preventive and cure for infectious and other diseases in ducks. Invaluable both for internal use and external application in cases requiring local treatment. If a bird shows signs of ill-health, treat at once with

IGLODINE
Antiseptic
8d., 1/-, 1/10½, 2/11

From all Chemists

Send 1d. Stamp for direction Booklet to:

The **IGLODINE Co. Ltd.**
Iglodine Buildings, Pilgrim Street
NEWCASTLE-UPON-TYNE

WANTED

DUCKS' FEATHERS

BEST POSSIBLE
. PRICES PAID .

CASH BY RETURN

Prices on application

SPITZ BROS.
Eagle Yard, High Road
TOTTENHAM, N.17

DOCTORS USE IT

CURICONES

for

RHEUMATISM *and* ALL KINDRED ILLS

FROM ALL CHEMISTS

Want to Buy?

- HORSES ?
- CATTLE ?
- SHEEP ?
- PIGS ?
- STOCK COCKERELS ?
- BREEDING HENS ?
- LAYING YEARLINGS ?
- LAYING PULLETS ?
- GROWING PULLETS ?
- DAY-OLD CHICKS ?
- HATCHING EGGS ?
- POULTRY EQUIPMENT ?
- AGRICULTURAL IMPLEMENTS ?
- ANY FARM REQUIREMENTS ?

THEN WRITE TO

The Best Farming Place in Eastern England, the Most Popular Place from which to buy.

LOWER BROS.

Farms and Branches under the control and management of Lower Bros. throughout Essex, Suffolk and Hertfordshire. Total 30,000 Acres.

Farmers, Stock-Breeders and Manufacturers

WILLOWS-GREEN - CHELMSFORD - ESSEX

Phone : GREAT LEIGHS 230 (Day and Night). Grams : "L.B. STOCKBREEDERS, GT. LEIGHS."

ADVERTISEMENTS

Books for Poultry Keepers

DOMESTIC POULTRY KEEPING (Fifth Edition). Elementary principles of poultry keeping—choice of breeds, accommodation, feeding and ailments, to egg preservation, etc. Fully illustrated. Price 1/-, by post 1/2.

NATURAL HATCHING AND REARING (Second Edition). By C. G. MAY. Fully illustrated. Price 1/-, by post 1/2.

BANTAMS FOR EGGS. By C. G. MAY. Housing—Management—Breeds and their Varieties—Feeding—Breeding and Rearing—Common Ailments. Price 1/-, by post 1/2.

CHICK MANAGEMENT—From the Day-Old Stage (Third Edition). By I. W. RHYS, N.D.P. Describes in detail all common brooding systems, including the hen, fold units, range and tier brooders. Feeding schedules, etc. Price 1/-, by post 1/2.

STARTING POULTRY KEEPING (Fifth Edition). By the Editorial Staff of "Poultry World" for war-time poultry keepers, with reference to emergency conditions. Fully illustrated. Price 1/6, by post 1/8.

POULTRY BREEDING AND PRODUCTION (Third Edition). By W. POWELL-OWEN. Covers every phase of breeding, including Waterfowl, Turkeys, Bantams and Guinea Fowl. 148 Pages. 68 Illustrations. Price 2/6, by post 2/8.

LAYING CAGES AND BATTERIES—Indoor and Outdoor. By W. POWELL-OWEN. The only publication on this intensive method of egg production. Separate supplement on War-time Feeding. Fully illustrated. Price 2/6, by post 2/8.

POULTRY AILMENTS. By W. P. BLOUNT, F.R.C.V.S. Suitable for both the beginner and the experienced. Glossary and Index. Cloth; Crown 8vo. 305 Pages—34 Illustrations in half-tone and line. Price 5/-, by post 5/5.

GEESE. Breeding, Rearing and General Management by REGINALD APPLEYARD. Illustrated. Price 2/6, by post 2/8.

PRACTICAL POULTRY KEEPING. Compiled by the advisory staff of "Poultry World." Covers every branch of poultry keeping. Of particular value to anyone contemplating poultry farming as a commercial undertaking. Fifteen chapters, 72 illustrations, Glossary and Index. 8vo. Price 5/-, by post 5/3.

GROW YOUR OWN EGGS or Silage from your Garden. By ARTHUR SMITH. Detailed instructions on conserving summer crops for winter feeding. Construction of a home-made drum silo, pit silo and barrel silo, suitable crops, filling instructions and information on feeding, etc. Price 6d., by post 7d.

EGG AND FOOD RECORD CARD. Provides for a daily record of eggs for twelve months, weekly food bills and an annual summary. Price 2d. each, by post 3d.

Obtainable through any newsagent, or direct by post from

The Publishers : **POULTRY WORLD**, Dorset House, Stamford St., S.E.1

Lightning Source UK Ltd.
Milton Keynes UK
UKHW040627290419
341788UK00001B/108/P